About Island Press

Since 1984, the nonprofit organization Island Press has been stimulating, shaping, and communicating ideas that are essential for solving environmental problems worldwide. With more than 800 titles in print and some 40 new releases each year, we are the nation's leading publisher on environmental issues. We identify innovative thinkers and emerging trends in the environmental field. We work with world-renowned experts and authors to develop cross-disciplinary solutions to environmental challenges.

Island Press designs and executes educational campaigns in conjunction with our authors to communicate their critical messages in print, in person, and online using the latest technologies, innovative programs, and the media. Our goal is to reach targeted audiences—scientists, policymakers, environmental advocates, urban planners, the media, and concerned citizens—with information that can be used to create the framework for long-term ecological health and human well-being.

Island Press gratefully acknowledges major support of our work by The Agua Fund, The Andrew W. Mellon Foundation, Betsy & Jesse Fink Foundation, The Bobolink Foundation, The Curtis and Edith Munson Foundation, Forrest C. and Frances H. Lattner Foundation, G.O. Forward Fund of the Saint Paul Foundation, Gordon and Betty Moore Foundation, The JPB Foundation, The Kresge Foundation, The Margaret A. Cargill Foundation, New Mexico Water Initiative, a project of Hanuman Foundation, The Overbrook Foundation, The S.D. Bechtel, Jr. Foundation, The Summit Charitable Foundation, Inc., V. Kann Rasmussen Foundation, The Wallace Alexander Gerbode Foundation, and other generous supporters.

The opinions expressed in this book are those of the author(s) and do not necessarily reflect the views of our supporters.

Great Plains Regional Technical Input Report

Great Plains Regional Technical Input Report

LEAD AUTHORS

Dennis S. Ojima
Jean Steiner
Shannon McNeely
Karen Cozetto
Amber N. Childress

ISLANDPRESS

Washington | Covelo | London

Printed on recycled, acid-free paper

Manufactured in the United States of America

10 9 8 7 6 5 4 3 2 1

Citation: Ojima, D.S., J. Steiner, S. McNeeley, K. Cozetto, A.N. Childress., A. Cole, J. Brown, G. Collins, L. Ferris, B. Gough, J. Gross, J. Hestbeck, D. Kluck, R. McMullen, J. Rattling Leaf, M. Shafer, M. Shulski, J. Yarbrough, M. Drummond, J. Morgan, T. Howell, S. Markstrom, H. Lazrus, K. Averyt, S. Skagens, K. Kunkel, L. Stevens, S. Stevens, M. Kruk, D. Thomas, E. Janssen, K. Hubbard, N. Umphlett, K. Robbins, L. Romolo, A. Akyuz, T. Pathak, T. Beragntino, E. Wood, K. Miller, B. Gascoigne, S. Tellinghouse, V. Tidwell, C. Aldridge, M. Rose, L. Wellings, T. Brown, J. Ramirez. 2015. *Great Plains Regional Technical Input Report.* Washington, DC: Island Press.

Keywords: climate change, Great Plains, National Climate Assessment, water resources, natural resources, agricultural system, conservation, land management, energy resources, human health, disease, adaption, mitigation, tribal landscapes, strategies

About This Series

This report is published as one of a series of technical inputs to the Third National Climate Assessment (NCA) report. The NCA is being conducted under the auspices of the Global Change Research Act of 1990, which requires a report to the President and Congress every four years on the status of climate change science and impacts. The NCA informs the nation about already observed changes, the current status of the climate, and anticipated trends for the future. The NCA report process integrates scientific information from multiple sources and sectors to highlight key findings and significant gaps in our knowledge. Findings from the NCA provide input to federal science priorities and are used by U.S. citizens, communities and businesses as they create more sustainable and environmentally sound plans for the nation's future.

In fall of 2011, the NCA requested technical input from a broad range of experts in academia, private industry, state and local governments, non-governmental organizations, professional societies, and impacted communities, with the intent of producing a better informed and more useful report. In particular, the eight NCA regions, as well as the Coastal and the Ocean biogeographical regions, were asked to contribute technical input reports highlighting past climate trends, projected climate change, and impacts to specific sectors in their regions. Each region established its own process for developing this technical input. The lead authors for related chapters in the Third NCA report, which will include a much shorter synthesis of climate change for each region, are using these technical input reports as important source material. By publishing this series of regional technical input reports, Island Press hopes to make this rich collection of information more widely available.

This series includes the following reports:

Climate Change and Pacific Islands: Indicators and Impacts
Coastal Impacts, Adaptation, and Vulnerabilities
Great Plains Regional Technical Input Report
Climate Change in the Midwest: A Synthesis Report for the National Climate Assessment
Climate Change in the Northeast: A Sourcebook
Climate Change in the Northwest: Implications for Landscapes, Waters, and Communities
Oceans and Marine Resources in a Changing Climate
Climate of the Southeast United States: Variability, Change, Impacts, and Vulnerability
Assessment of Climate Change in the Southwest United States
Climate Change and Infrastructure, Urban Systems, and Vulnerabilities: Technical Report for the US Department of Energy in Support of the National Climate Assessment
Climate Change and Energy Supply and Use: Technical Report for the US Department of Energy in Support of the National Climate Assessment

Electronic copies of all reports can be accessed on the Climate Adaptation Knowledge Exchange (CAKE) website at *www.cakex.org/NCAreports*. Printed copies are available for sale on the Island Press website at *www.islandpress.org/NCAreports*.

Contents

Acknowledgments

The convening lead authors for this report were Dennis S. Ojima, Colorado State University, and Jean Steiner, USDA-ARS. We thank the following partners for their invaluable contributions, insights, and feedback:

Editorial Team

Dennis S. Ojima (Colorado State University), Jean Steiner (USDA-ARS), Shannon McNeeley (NCAR), Amber N. Childress (Colorado State University), Karen Cozetto (University of Colorado, Boulder)

Lead Authors

Andy Cole (USDA-ARS), Joel Brown (USDA NRCS), Gary Collins, Laura Ferris (EPA), Bob Gough (Intertribal COUP), John Gross (NPS), Jay Hestbeck (USGS), Doug Kluck (NOAA), Ryan McMullen (USDA RD), James Rattling Leaf, Mark Shafer (SCIPP RISA, OK), Martha Schulski (HPCC, UNL), James Yarbrough (EPA)

Contributing Authors

Mark Drummond, Jack Morgan, Terry Howell, Steve Markstrom, Heather Lazrus, Kristen Averyt, Susan Skagens, Ken Kunkel, Laura Stevens, Scott Stevens, Michael Kruk, Devin Thomas, Emily Janssen, Kenneth Hubbard, Natalie Umphlett, Kevin Robbins, Luigi Romolo, Adnan Akyuz, Tapan Pathak, Tony Bergantino, Elizabeth Wood, Kathy Miller, Billy Gascoigne, Stacy Tellinghouse, Vince Tidwell, Cam Aldridge, Matt Rose, Leigh Wellings, Tom Brown, Jorge Ramirez

Copy Editor

Joshua Zaffos

Support provided by

Support for the Great Plains assessment technical team was provided by Colorado State University, Department of Interior North Central Climate Science Center and the United States Geological Survey, United States Department of Agriculture-Agricultural Research Service. Additional support came from the National Park Service, USDA Natural Resources Conservation Service, US Forest Service, US Environmental Protection Agency Region 8, National Ocean Atmospheric Administration, Western Water Assessment Regional Integrated Sciences and Assessment Center (RISA), Southern Climate Impacts Planning Program RISA, Colorado State University, University of Colorado-Boulder, University of Nebraska-Lincoln, National Center for Atmospheric Research, and the Intertribal Council On Utility Policy.

Acronyms

ASCE	American Society of Civil Engineers
CCC	Canadian Climate Center
CCS	carbon capture and sequestration
CO_2	carbon dioxide
COOP	Cooperative Observer Network
CRP	Conservation Reserve Program
CSC	DOI Regional Climate Science Centers
EAP	Emergency Action Plan
EPA	Environmental Protection Agency
EWe	ethanol
FEMA	Federal Emergency Management Agency
GHG	Greenhouse gas
HAD	Hadley Center
N_2O	Nitrous oxide
NARCCAP	North American Regional Climate Change Assessment Report
NCDC	National Climate Data Center
NFIP	National Flood Insurance Program
NGO	nongovernmental organizations
NOAA	National Oceanic and Atmospheric Administration
NRCS	National Resources Conservation Service
NWS	National Weather Service
OTSA	Oklahoma Tribal Statistical Area
PET	potential evapotranspiration
RISA	Regional Integrated Science and Assessment

RPS	Renewable Portfolio Standard
TCW	total consumptive water
UHI	urban heat island
USDA	U.S. Department of Agriculture
WGA	Western Governors' Association
WSWC	Western States Water Council

Executive Summary

Great Plains Systems and Climate Change assesses how the Great Plains social-ecological system has been shaped by changing climate conditions and how future projections of climate change will result in a need for further adaptation and preparedness. This effort is part of the 2014 United States Global Change Research Program National Climate Assessment as required by the United States Congress.

The Great Plains region plays a very important role in providing food and energy to the economy of the United States from the great corn and wheat fields and rangelands in the agricultural sector, the Bakken Shale formation in North Dakota, the abundant coal and coal bed methane in the Wyoming and Montana Powder River Basin, bioenergy and wind farms in Texas in the energy sector. This makes the economy and livelihoods in the region extremely sensitive to climate, which means big implications of climate change impacts on the Great Plains region as well as mitigation strategies to reduce greenhouse gases critically important for the entire country. The region is also the home to 65 registered Native American tribes who stand to be vulnerable to climate change while also potentially contributing to innovation in sustainable practices and an alternative energy future. This all makes the Great Plains a complex and interesting place to look at the impacts of climate variability and change.

The Great Plains region is characterized by both high spatial and high temporal climate variability, however, throughout the region **climate change is already happening in the Great Plains** with an overall warming trend over the last 20 years both annually and in the summer. Climate change is being experienced in a variety of ways such as increased night-time temperature, increased intensity of extreme precipitation events, extended growing season, extended severe droughts, and elevated atmospheric CO_2 concentrations. Climate change is projected to continue into the future with more extreme heat events, droughts, and floods. Expected impacts include decreased water availability and increased competition for uses, changed water quality, expansion of weeds, pests, and diseases, changes to plant-animal communities and species composition, altered fire and storm patterns, and tree mortality, among others. Combined with changes in land use and land management, socio-economic and demographic changes, and uncertainty of our energy future, climate change will have substantial impacts on the ability to sustain natural resources, livelihoods, and well being in the Great Plains.

Over the last decade the region has seen significant extremes in climate and weather events from flooding in the Missouri River Basin, to exceptional drought in the Southern Plains, to fires and tornadoes resulting in billions of dollars in economic damage, morbidity, and mortality. Some of this unusual weather is the result of normal climate variability, but many climate experts understand these extremes as indicators of emerging climate changes, if not already a signal that we are seeing effects of a warming planet.

Key Findings

Multiple climatic and non-climatic stressors, of which climate change is one among many, put multiple sectors, livelihoods and communities at risk. The most vulnerable in the region are agriculture, water, ecosystems and rural and tribal communities.

The Great Plains climate is warming, and as of 2011, eight of the last ten summers have been above average temperature. Climate observations in the Great Plains show the warmest years on record were a tie between 1934 and 2006. The Northern Great Plains has experienced the most significant warming where North Dakota, for example, has experienced an annual average temperature increase of 0.26ºF per decade during the last 130 years, the fastest increase in the nation. Growing season has extended with first freeze in the fall coming later and last freeze in the spring coming earlier; the average growing season is longer by about 6 days from 1991-2010 compared to 1961-1990. Annual precipitation was greater than normal during the 1990s, less than normal during the early 2000s when most of the western U.S. experienced severe drought, and greater than normal in the last few years. However, while some areas experienced major flooding such as in the Northern Great Plains, other areas were simultaneously experiencing extreme drought conditions such as in the Southern Great Plains. This shows that climate change will not manifest in a uniform way across the Great Plains and that preparations must be made within geographic sub-regions to deal with a range of extreme weather and climate conditions.

Extreme hot temperatures will increase and climate projections show that temperature increases will be the largest in the summertime, which has huge implications for more heat waves, energy and water demands, and water scarcity. Mean summer temperature increases are projected to be 3.3ºF in 2035, 5.4ºF in 2055, and 9ºF in 2085. The number of days above 100ºF are projected to increase by 15 days by mid-Century with heat wave events nearly doubling in length.

Climate change impacts to the hydrological cycles will be felt throughout the region and across all sectors. Changes in precipitation patterns, the timing of seasonality of rain and snow and the alterations of large scale circulation patterns have major impacts on water availability in the region. Decreased snowfall in lower mountain elevations combined with earlier snow melt and earlier spring runoff will have big impacts on the timing and amount of streamflow affecting irrigators and other diverters and users of suface water resources. This can also affect lead to diminished late season streamflow, which impacts fish and riparian ecosystem health as well as the ability for late summer/early fall irrigation of crops. Increased conflict between competing users of water is likely to increase within and between states.

- **The Northern Great Plains region is expected to increase in extreme precipitation events, leading to more damaging flooding in some areas.** Heavy precipitation events could increase as high as 30% in some areas, such as South Dakota where flood disasters have been considerable over the last decade.
- **Drought is expected to increase, especially in the already drier western portions of the Great Plains.** The 2011 drought in the Southern Plains is consistent with climate projections, however, it is also strongly associated with natural variability and specifically the La Niña phase of El Niño Southern Oscillation.

Both climate variability and climate change increase drought risk for the region, and while some efforts are already underway as discussed in Chapters 8 and 9 of this report, much more effort is needed for assessment of drought vulnerability and for preparing for the impacts of extended severe droughts that are expected in the future.

- **Groundwater that is already stressed in the Great Plains, such as in the High Plains (Ogllala) Aquifer region will be exacerbated by climate change.** A combination of increased groundwater pumping, diminished water quality, drought, salinization, increased temperatures and evapotranspiration, and a variety of interacting surface and sub-surface dynamics all threaten the sustainability of groundwater in the region, all of which will likely be exacerbated by the effects of climate change.

Crops will be impacted differently across the Great Plains where some areas will benefit from a longer growing season and more rainfall where others parts will experience decreased productivity because of drought and extreme temperatures. Extended growing season could provide more options for crop diversification.

Bioenergy production is an area of potential economic development in the United States, especially in rural areas, as well as a potential source to contribute to domestic energy independence and reduction of fossil fuel use and greenhouse gas emissions, however, the limits in the Great Plains must be carefully considered. One limitation on the expansion of corn ethanol production in the Great Plains is the use of ground water in already vulnerable and water-stressed areas. Careful consideration must be given to producing corn ethanol in areas that are not already at high risk for water stress. Several states in the Great Plains region are being threatened by water shortages across local, state, and regional scale, yet, there is an economic incentive, and pressure to grow corn for energy in the High Plains is strong where irrigation costs are comparatively small to the amount of increased production it results in. Competition between water for biofuels and other demands will be highest in the nation in the High Plains region. These regional tradeoffs that must be considered between fossil fuel energy, renewable energy to meet Renewable Portfolio Standards (RPSs), and the water necessary to meet these economic and environmental goals. It is critical to take regional and local context into account for policy and planning across all scales of governance; and it requires careful planning at the watershed level within and between states at a regional level. Some experts have suggested "next generation" biofeed stock such as perennial grasses and woody biomass will help meet the needs for bioenergy, however, the extent to which this potential exists or is limited by local and regional conditions in the Great Plains has yet to be determined.

The many Native American tribes throughout the Great Plains are located in relatively marginal areas lacking access to fertile soils, appropriate housing, electricity and energy sources, food and water sources. All of this makes many tribal members highly vulnerable to the impacts of climate change. However, many tribal governments have started the process of developing and sustaining viable economies on their lands and providing a set of strategies to cope with climate change.

The population of the Great Plains has been moving into urban areas, which will also see many impacts from climate change. Increased temperatures will lead to more heat wave events and increased morbidity and mortality of the highly vulnerable populations in cities. Heat waves with long stretches over 100ºF will increase leading to problems with transportation infrastructure as well as loads on electricity systems, highly dependent on water for cooling. When heat events combine with droughts this can be disastrous for cities when heat and water scarcity threaten power outages, water use curtailment, and heavy damages to crops and livestock as Texas has experienced in 2011. Climate change will also bring intense storms taxing already stressed stormwater and sewage infrastructure and water quality.

The nexus between water and energy is increasingly important to understand as climate change complicates this already complex relationship. Water is needed for energy extraction, production, and power; energy is needed to move and treat water. Mitigation and adaptation strategies must take into consideration the water-energy nexus as choices about our energy future require tradeoffs for the use of land and water; and adaptation choices must be made in the context of large uncertainties about the availability of both. Because of the importance of energy and water in the Great Plains, the water-energy nexus is critical to understand.

Ecosystems are already stressed by climate variability and change such as droughts, floods, and winter storms that have altered plant community phenology, hydrological dynamics of streamflow and wetland dynamics. Warming water temperatures are already pushing aquatic species to their limits, and combined with multiple stressors such as impoundments, diversions, sedimentation, decreased water quality, and the changing of the timing and amount of hydrological events critical to breeding or migration times will be exacerbated by climate change. Some species will be pushed past their threshold limits as a result.

Climate change will shift the geographic distribution of diseases in the Great Plains, which will affect human, ecosystem, and livestock health.

Multiple coping response and adaptation strategies are already being implemented by state and federal agencies, urban areas, tribes, and natural resource managers in decision makers in the Great Plains. However, there must be continued support across all levels of government as well as the private sector and industry to adequately mitigate and adapt to climate change.

Research needs

Effectively addressing climate change and its effects on ecosystems, resources, and society will require coordination in the research and observation capabilities of multiple organizations, institutions, and government programs. This report outlines a rich and varied set of activities to address climate change, however, it is apparent that there are few well-coordinated efforts between agencies or institutions. There is an additional challenge of knowledge dissemination and monitoring of impacts, vulnerabilities, and adaptations or coping strategies for responding to or preparing for climate change. This lack of coordination and communication results in a great inefficiency and ability to assess climate change impacts to focus research activities more strategically in the region. However, some good examples are emerging such as the initiatives of the Western Governors Association (WGA), the Western Federal Agency Support Team (discussed in Chapter 9). Wherever possible participatory research, iterative risk-based analysis between researchers and stakeholders, natural resource managers, policy makers are needed for collaborative decision making to deal with the impacts of climate change.

Research efforts have brought attention to the role ecosystems have in providing key economic goods and the ecosystem services that sustain, regulate, and support life on Earth (Costanza et al. 1997, Daily 1997, Daily and Ellison 2002). But the societal and ecological contributions of the 'underpinning' services provided by ecosystems often remain 'invisible' and unvalued (or undervalued). The array of such services is broad, from those services that regulate critical human-environment processes (e.g., climate, disease, flooding, detoxification) to services that support economic activity (e.g., soil formation, primary productivity, nutrient cycling, pest control, pollination).

Incorporating ecosystem services into the decision making process allow managers to better understand effects of land use and management. Development of a more robust adaptive management approach that recognizes the existence of transitioning landscapes (e.g., shrub invasion of grasslands) and the importance of change as a basic component of the system dynamic (Berkes and Folke 1998; Gunderson and Holling 2002; Walker and Meyers 2004; Tschakert et al. 2007) needs further attention.

Forecasting technologies have advanced tremendously over the past decade; incorporating field observations, remote sensing, and modeling systems to provide seasonal forecasts of crop and ecosystem productivity. In addition, observations of ecosystem indicators associated with biotic assemblages (e.g., host-pest relationships), ecosystem functions (e.g., water use efficiency, nutrient cycling, and carbon storage and fluxes), and structural changes (e.g., woody to herbaceous ratio, bare soil exposure) can provide clues to emerging dangerous thresholds. Complex multi-dimensionality of ecological thresholds can be resolved through the use of integrative modeling and analytical approaches. Development of a threshold prognostic or ecological forecasting tool to evaluate probabilities of achieving a threshold event would be extremely helpful in managing natural resources and developing adaptive management strategies to maintain ecosystem services. A number of ecosystem services can also be monitored to assess the impacts of change to society and vice versa.

A systems approach which incorporates aspects of ecosystem services, livelihoods, and ecosystem integrity needs further development. A number of approaches on social-ecological vulnerabilities have been developed over the past decade or more (Adger et al., 2001, Moss et al., 2000; Turner et al. 2003 (Ford et al. 2010, Adger et al. 2007, Adger et al. 2004, Fussel and Klein 2006, Smit and Wandel 2006). Such approaches allow for better integration of environmental and societal metrics and variables to evaluate social-ecological vulnerability. These analyses would provide greater insight to the range of coping choices to make under various levels of adaptive capacity.

Adaptive capacity is constrained by factors that restrict people's set of options to choose from when environmental and social conditions change (Berkes and Folke 1998, Gunderson and Holling 2002). Institutional hierarchies often constrain adaptive mechanisms operating at the community level, determining in part how adaptation to climate change manifests through policy processes (Adger and Kelly 1999). Institutional responses to climate change are often best suited for mitigation of emergency situations and isolated events, rather than for slower onset, cumulative or systemic climate-related problems leading to disruption of social and ecosystem services. Institutional and regulatory entities are less well-suited to working with underlying social factors that determine vulnerability (Handmer et al. 1999). Where institutional rule-making occurs in a compartmentalized and fragmented framework, responses to climate change have been either nonexistent in the worst case, or case-based mitigation in the best case (McNeeley 2011).

Both scientists and managers suggested solutions or products that could reduce management barriers and improve climate change response. The reinforcement of partnerships was a common theme that was promoted through the workshop. Managers noted the need for a centralized mechanism to communicate current and ongoing research projects in the region. Products that promote education and awareness of local climate change issues, including additional webinars and workshops, were seen as critical to engage stakeholders and inspire action. Climate change can also present opportunities; for example, carbon sequestration can be a driver for implementing grassland restoration projects. Other participants suggested more specific measures, such as implementing changes in breeds or species of grazers to cope with changes in forage productivity or composition. New technologies can be integrated into management, such as the transmission of real-time remote-sensing data through wireless devices to better inform day-to-day management decisions, or use of social networking to bring together stakeholders.

There are lessons to be learned from efforts of the WGA that illustrate how state, federal, tribal, and academic communities can work in a coordinated fashion to develop and implement strategies to deal with critical regional needs related to climate change. Issues, including water resources, land use, forest fire, and conservation needs, have been proactively addressed over the years, as mentioned above. The WGA has helped to define issues and to provide a framework to address these across the West. Other regional efforts include river basin initiatives, such as the Missouri River Basin efforts and the various agency coordination efforts to deal with flood control, land use practices, and conservation efforts. These bodies have a goal to provide better communication and, where needed, coordination of actions to deal with specific issues. The energy sector also has regional action groups, as mentioned in previous sections of this report. However, assessments of climate change impacts and long-lasting climate change solutions need to be developed across sectors and include multiple stakeholders. We need to create a platform to support this more integrative effort in the research and the management activities implemented across the Great Plains.

SECTION 1

Great Plains

Scope, History and Recent Trends

The United States Great Plains stretches across a massive expanse of land, which has historically faced frequent extreme weather, limited water availability, and other ecosystem constraints that together define the region's environmental and social systems.

The variety of natural resources has helped make the region a main source of crops and livestock, food and fiber, for the nation. The same mix of factors creates natural diversity among wetlands and grasslands, which offer critical habitat for aquatic and terrestrial species. People living in the region have faced the challenges – and benefits – of living on the plains, dating back to Native American societies. Even though the Great Plains are characterized by climate variability, the rise in greenhouse gases and continued and projected climate changes will impact ecosystems, conservation efforts, economic and agricultural activities and, ultimately, human development in the region for the next century and beyond.

© Danny Martin www.dannymartinphotography.com

Chapter 1

Great Plains

Social-Ecological Setting (Climate-Environment-Society) Natural Resources and Wildlife Aspects

The Great Plains lie west of the Mississippi River and east of the Rocky Mountains, rising gradually from about 98 ft (30 m) above sea level to over 5,000 ft (1,524 m) at the foot of the mountains. Before their widespread conversion to intensive agriculture, the Great Plains were noted for their extensive grasslands, from tall-grass prairie in the east to short-grass prairie in the western High Plains. The generally low relief of the plains is broken in several places, notably by the Ozark and the Ouachita Mountains, which form the Interior Highlands, the only major mountainous region between the Rocky Mountains and the Appalachian Mountains (see Havstad et al. 2009).

The social-ecological setting of the Great Plains is comprised of a suite of sectors and communities which include ranchers, farmers, city dwellers, business entrepreneurs, energy developers, natural resource managers, recreationists, Native American tribes and others. These communities are connected to the abundant ecosystem services and natural resources in the region and are influenced by climate and weather patterns. Natural resources, such as rivers, rich soils, biodiversity, wildlife, and vegetation, of the Great Plains are sensitive to climate and weather patterns across the region. In addition, current market forces, policy and regulatory statutes, cultural trends, and jurisdictional and institutional structures affect management decisions and responses to changing social-environmental conditions in the region.

The US Great Plains cover an area of over 500,000 square miles (1.3 million km^2) in the Central US, which was historically a grassland landscape. The Great Plains cover all or part of 10 states in the central portion of the United States, including Montana, North Dakota, Wyoming, South Dakota, Colorado, Nebraska, Kansas, New Mexico, Oklahoma, and Texas. This assessment covers primarily a 9 state region which excludes New Mexico since it has been included more extensively in the Southwest Climate Assessment. The Plains are characterized by a temperature gradient that gets warmer from north to south and a precipitation gradient that gets wetter from west to east. Annual average precipitation ranges from 8 in (200 mm) in the west to approximately 43 in (1100 mm) in the east and southeastern portion of the region, and can be highly variable from year to year. There is strong seasonality and high variability in temperature and precipitation patterns. In addition, extreme weather events in the Great Plains, including droughts, floods, tornadoes, hail, ice storms, heat waves, blizzards, and, along the Texas Gulf Coast, hurricanes, occur.

In order to sustain ecosystem services, natural resources, and livelihoods in this diverse and variable climate environment greater understanding of environmental changes is required. Seasonality is an important factor affecting land systems, from agriculture; to energy sectors; to water, land, and forest management in the Great Plains. A change in the statistical mean for temperature or precipitation is not as important to these land uses as changes in variability or seasonal patterns of weather conditions. For example, hotter temperatures and less moisture during the growing season may impact range or crop production dramatically. When looking strictly at range and livestock systems, a number of potential impacts have been identified (Ojima and Lackett 2002, Morgan et al. 2011). First, forage production and quality will certainly be altered. Some of the changes may be beneficial, such as enhanced production under elevated CO_2, while other changes may be deleterious, such as the fact that the forage may be less nutritious. Carrying capacity will be impacted and there will be shifts in vegetative communities. Increased extreme events may lead to poor performance of existing livestock breeds of intensive livestock systems if the current breeds have thermal thresholds which are exceeded (Hahn et al. 1998).

The Great Plains' communities, though residing in a bountiful environment, are sensitive to changes in markets, weather, water availability, and policies which affect multiple factors determining the outcomes of their livelihoods.

Great Plains Communities

NATIVE AMERICAN LEGACY

Native Americans in the Great Plains have a rich and varied history, extending back many generations. Nomadic and semi-nomadic Native American populations occupied the Great Plains, prior to European incursion. These nomadic Plains tribes followed seasonal migrations of vast herds of buffalo and other wildlife. Other Plains' tribes were semi-sedentary, not only hunting buffalo but also living in villages and raising crops. The tribes were very successful at adapting to natural cycles and weather extremes (Maynard 1998). Although they altered ecosystem dynamics in their own way, the hunting-gathering societies characterizing Native American cultures did not extensively alter the flow of water and nutrients in the ecosystems of the Great Plains. The lands presently occupied by most tribes are located in relatively marginal areas. Many tribes thus lack access to fertile soils. They often also lack access to traditional energy, and food and water resources, which make living in tribal areas more challenging. However, tribal governments have started the process of developing and sustaining viable economies on their lands and providing a set of strategies to cope with climate change.

Today, the Great Plains are home to 65 Native American tribes. According to the 2000 Census (U.S. Census Bureau 2000), about 20% of all American Indian and Alaska Natives call the Great Plains home and, close to 450,000 of them live on Great Plains reservations or Oklahoma Tribal Statistical Areas. In Oklahoma and South Dakota, 11.5% and 9.0% of the state populations, respectively, claim at least part American Indian and Alaska Natives ancestry. According to the 2000 census, on-reservation/ or Oklahoma Tribal Statistical Areas Native American unemployment rates in the Great Plains were almost two times the national average, and in certain states including Nebraska, South Dakota, and Wyoming the rates were roughly four times the national average.

The median household income on-reservation/ or Oklahoma Tribal Statistical Areas was about $26,700, which was roughly 36% below the national average of $42,000. In North Dakota, South Dakota, Wyoming, and Texas; the median household income are about 90% below the national average.

These economic hardships and the lack of well-paying, long-term livelihoods cause Native populations to be more vulnerable to climate change impacts than wealthier sectors of society. The resources of many tribal governments are already extremely constrained without any additional climate stress (Maynard 1998).

RURAL COMMUNITIES AND LIVELIHOODS

Settlement of the Great Plains proceeded rapidly after laws such as the Homestead Act of 1862 were passed, allowing settlers to own 160 acres (65 hectares) of land after five years of residency. The population in the Great Plains grew steadily until 1930, when the Dust Bowl period began (mid-1930s). After the 1930s, wheat cultivation rebounded from the effects of the Dust Bowl as war demands for food increased. Since the late 1930s, farms and ranches in the Great Plains have been decreasing in number and increasing in size (Lackett and Galvin 2008). Expansion of farms in the Great Plains may be partially explained by the low and uncertain precipitation, leading to low per-acre yields and increased acreages in order to increase incomes, in addition to consolidation of land holding due to economic and technological changes.

The region's socio-economic system is characterized by extensive rural livelihoods with a recent concentration of populations into urban areas. As of 2010, there were almost 42 million people (approximately 13% of the total US population) living in the 9 US Great Plains states, including Colorado (USDA Economic Research Service 2012a). The average population density over the region is about 66 people per square mile, with a median of 10 people per square mile (U.S. Census Bureau 2010). Although the region's population has been increasing, the growth has not been equitable across counties. Urban population numbers have grown to almost 33 million persons in the past 20 years (U.S. Census Bureau 2010). Thirty-nine percent of the counties in the Great Plains have declined in population from 1990 to 2010 (U.S. Census Bureau 2010), with rural counties much more likely to lose population than those with some urban development.

Although the vast majority of the Great Plains landscape consists of remote areas, nearly 80% of the almost 42,000,000 residents of the region live in urban areas (U.S. Census Bureau 2010). As with almost all other facets of life in the region, there is great diversity in population density and socio-economics, even within rural and urban areas. The gradient in population density ranges from farming communities near metro areas in the Southern Great Plains to communities centrally focused on livestock grazing in the North (Figure 1.1). Rural areas range from counties in Montana with less than 1 person per every 2 square miles (5 km^2) to counties outlying major metro areas, that will likely transition to urban areas in coming years. The degree of urban residents varies from state to state ranging from South Dakota, which is almost evenly split between urban and rural residents, to Colorado and Texas, each of which has over 80% of their residents living in urban areas (U.S. Census Bureau 2010). Texas contains two of the most populous and fastest growing metropolitan areas in the country, Dallas-Fort

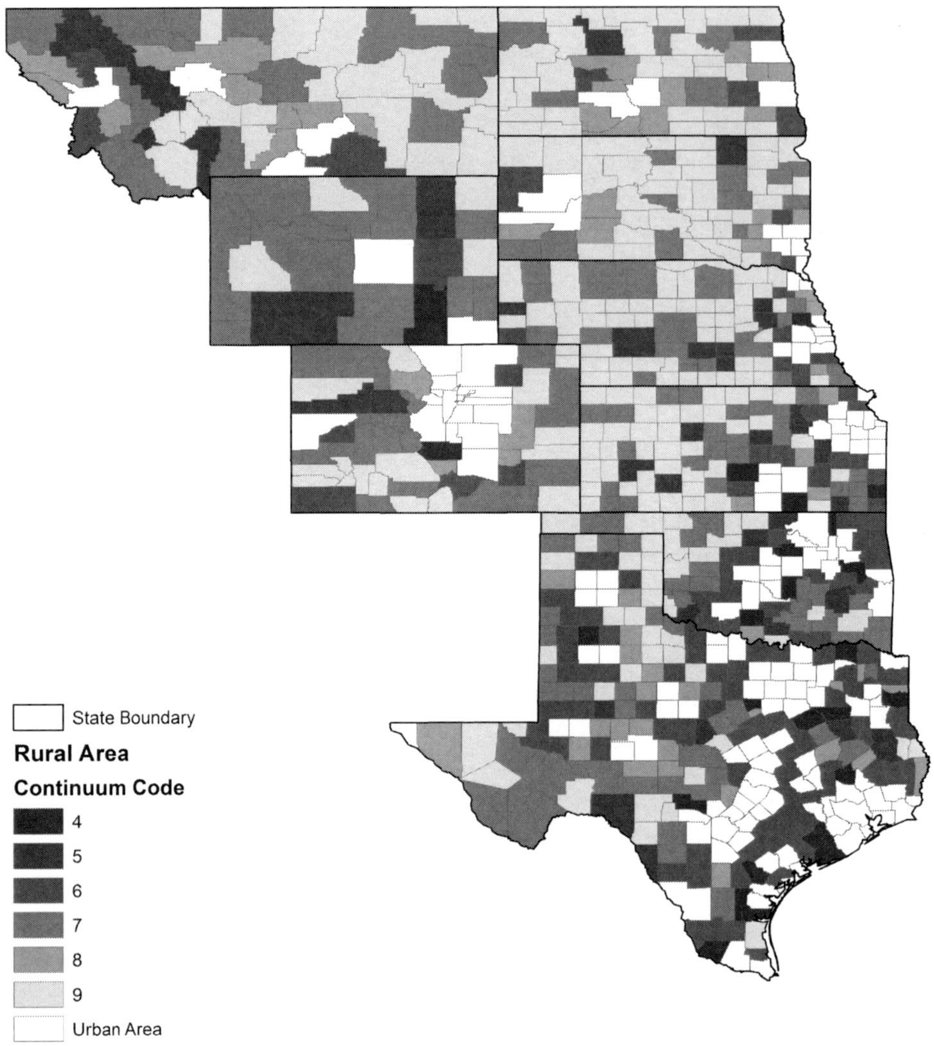

Figure 1.1. Great Plains Rural and Urban Counties (USDA Economic Research Service 2012a).

Worth-Arlington and Houston-Sugar Land-Baytown, which as of the 2010 Census rank fourth and sixth in the nation in terms of population magnitudes (U.S. Census Bureau 2010, Mackun and Wilson 2011).

CHARACTERISTICS OF GREAT PLAINS RURAL AREAS

Although rural areas in the Great Plains are diverse, there are some defining characteristics of counties within rural areas. Potentially the most significant common feature of rural counties is their economic dependence on agriculture. Forty-five percent of nonmetro counties are farm-dependent, compared to just four percent of metro counties (Figure 1.2) (USDA Economic Research Service 2012b). Farm dependence, as defined by the U.S. Department of Agriculture (USDA), is based on two thresholds. Farm earnings must account for an annual average of at least 15 percent of total county earnings,

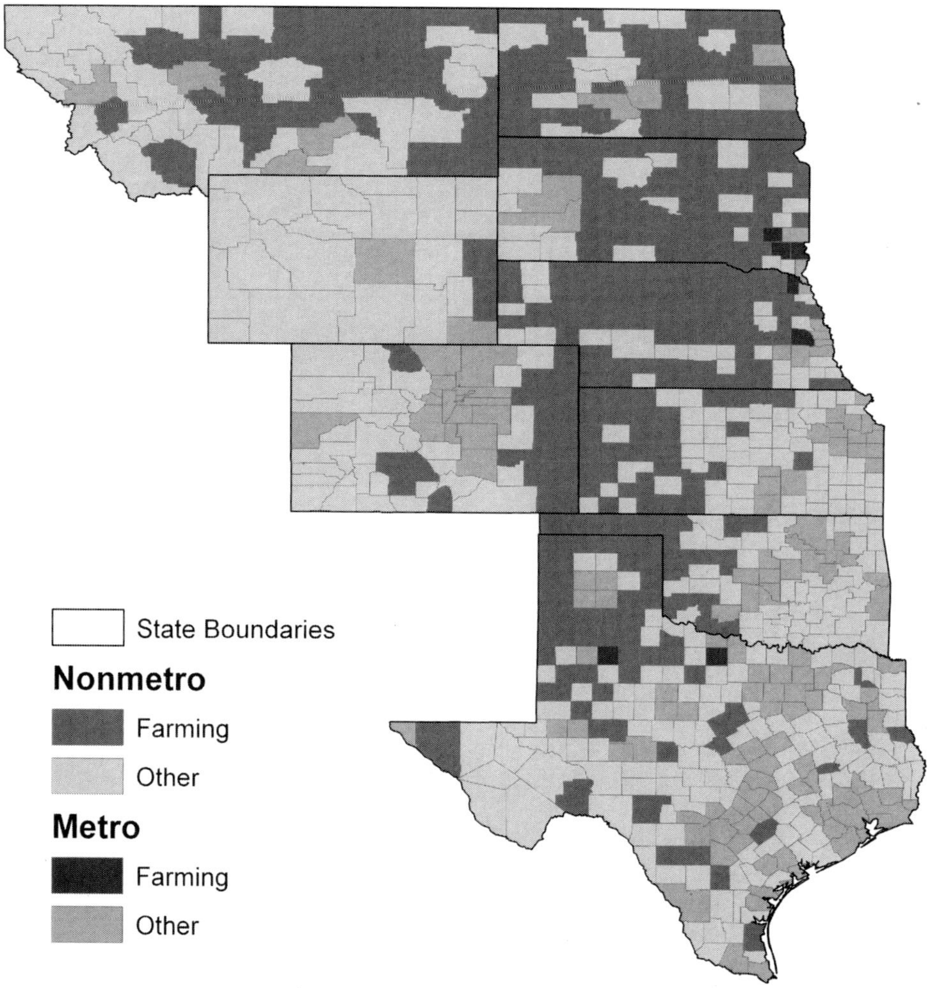

Figure 1.2. Great Plains Farm-Dependent Counties (USDA Economic Research Service 2012a)

or farm occupations must account for at least 15 percent of all occupations of employed county residents. However in some regions of the Great Plains, recent energy development associated with gas extraction has created an economic surge.

Due to the remoteness of rural areas, many residents lack easy access to resources and services. The scope varies as communities become more remote and access to resources and services generally decreases. Analysis of water availability and sanitation services by the Rural Community Assistance Partnership determined that the percentage of households lacking proper water and sanitation is highest in places with populations of less than 1,000 and rural farm populations. This is largely attributed to rural areas lacking economies of scale to support such services without subsidization and a lack of financing or technical assistance (Vaswani and Gasteyer 2004).

For the same reason, rural areas typically have less access to other public services, such as medical care, fire departments, and schools. These services decline as populations get smaller and counties become more remote from metro areas. Accessibility of

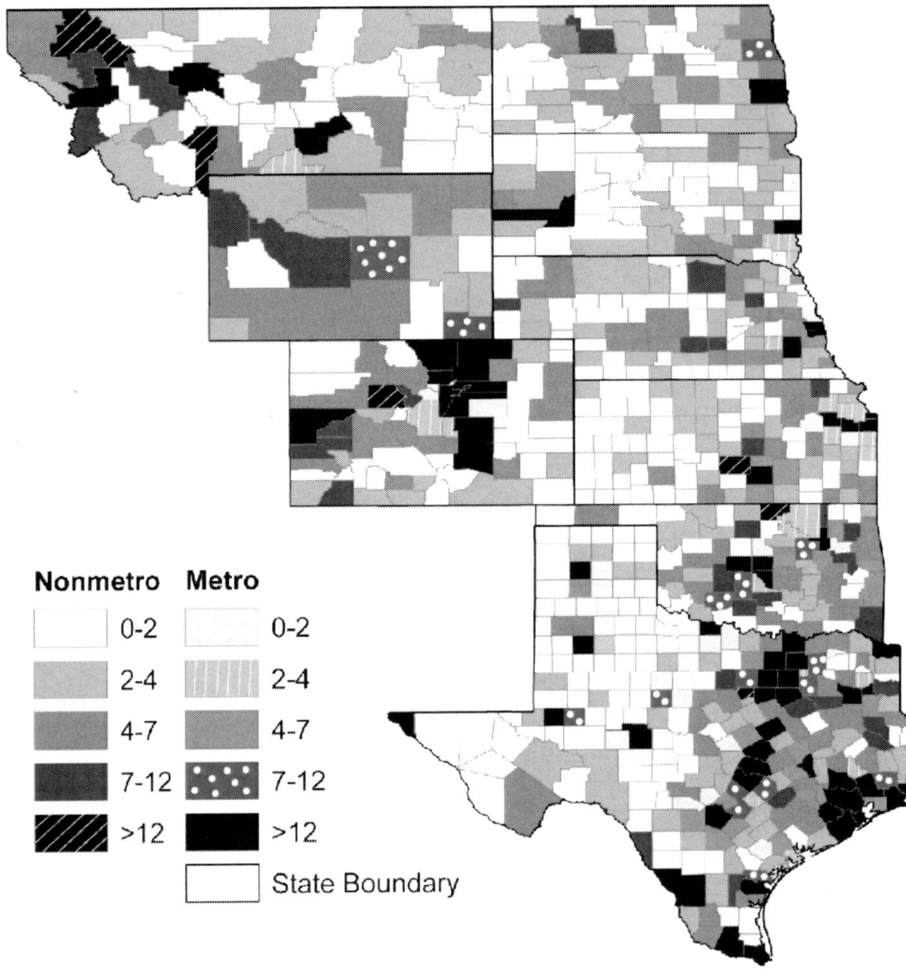

Figure 1.3. Number of Grocery Stores per County (USDA Economic Research Service 2012b)

health care tends to deteriorate as geographic isolation increases and population density declines (Lal et al. 2011). Emergency response systems are often less effective due to the population dispersion and geographic isolation. Lal et al., (2011) concludes that the combined effects of changing demographics and increasing health costs are more likely to make it difficult to supply rural areas with adequate public health services.

Access to goods is also marginal in rural, isolated areas. Throughout the Great Plains, grocery stores are many miles apart in most non-metro counties, causing residents to drive long distances for food (Figure 1.3). Similar to medical care, this results in families spending a larger proportion of income on food than urban residents (USDA Economic Research Service 2012c).

Overall, average income in rural counties is lower, than urban areas. As a result people who receive college degrees typically do not return and fewer people with degrees are employed in the rural labor market (Lal et al. 2011). A comparison by Lal et al., (2011) of nationwide rural-urban dynamics determined that the widening rural-urban income

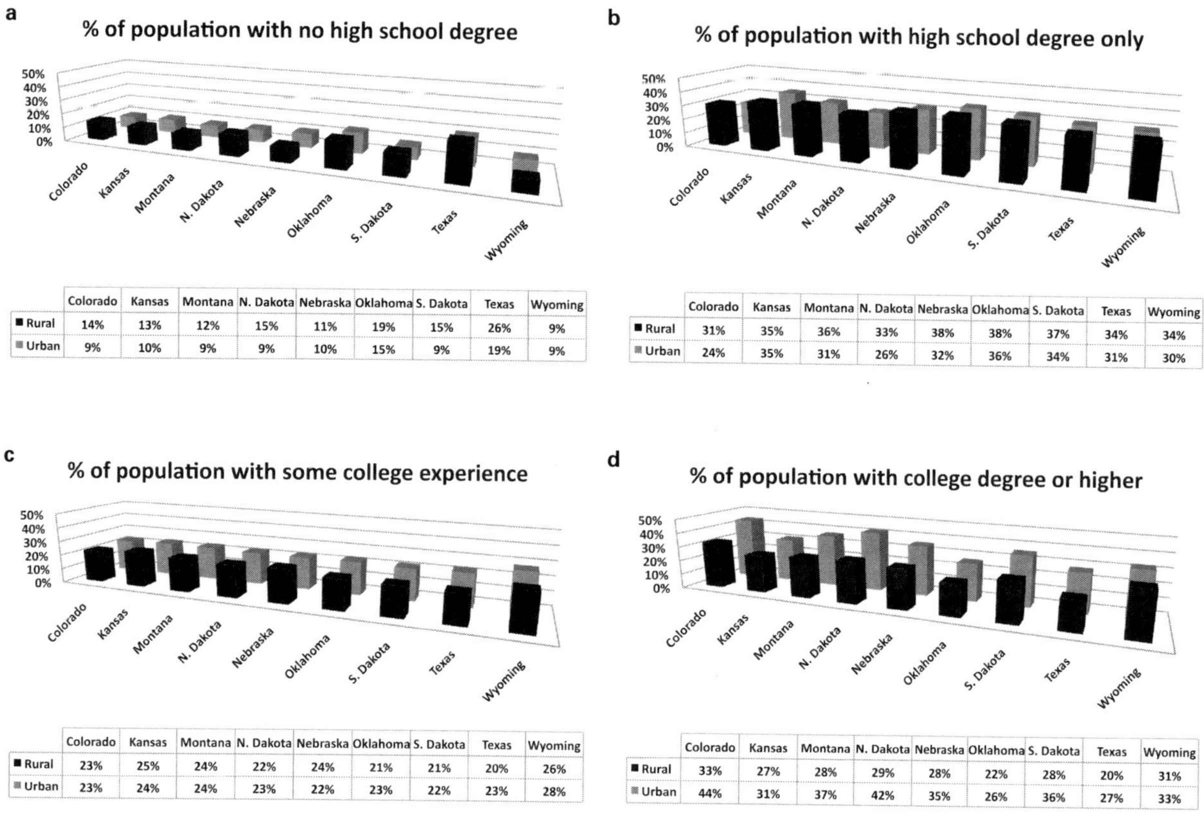

a

	Colorado	Kansas	Montana	N. Dakota	Nebraska	Oklahoma	S. Dakota	Texas	Wyoming
■ Rural	14%	13%	12%	15%	11%	19%	15%	26%	9%
■ Urban	9%	10%	9%	9%	10%	15%	9%	19%	9%

b

	Colorado	Kansas	Montana	N. Dakota	Nebraska	Oklahoma	S. Dakota	Texas	Wyoming
■ Rural	31%	35%	36%	33%	38%	38%	37%	34%	34%
■ Urban	24%	35%	31%	26%	32%	36%	34%	31%	30%

c

	Colorado	Kansas	Montana	N. Dakota	Nebraska	Oklahoma	S. Dakota	Texas	Wyoming
■ Rural	23%	25%	24%	22%	24%	21%	21%	20%	26%
■ Urban	23%	24%	24%	23%	22%	23%	22%	23%	28%

d

	Colorado	Kansas	Montana	N. Dakota	Nebraska	Oklahoma	S. Dakota	Texas	Wyoming
■ Rural	33%	27%	28%	29%	28%	22%	28%	20%	31%
■ Urban	44%	31%	37%	42%	35%	26%	36%	27%	33%

Figure 1.4a–d. Education Graphs (U.S. Census Bureau 2010a)

gap is associated with lower costs of living in rural areas, lower educational attainment, less competition for workers among employers and fewer highly skilled jobs. This trend can also be seen across the Great Plains region, expressed in trends of lower educational attainment in rural areas (Figure 1.4).

The lack of skilled jobs in rural communities has led to an out-migration of working-age populations from agricultural communities (Figure 1.5). Parton et al. (2007) conclude that this out-migration of youth has had the secondary consequence of reducing fertility and intensifying the downsizing of many aspects of community life, particularly activities and schools that focus on children, leading to the acceleration of further out-migration. Counties with higher levels of irrigated agriculture tend to see lower rates of out-migration and have steadier and relatively younger populations. It has yet to be determined whether this trend will sustain over time or if it is just slower because of improved economic conditions (Parton et al. 2007).

This chronic out-migration and lower fertility rates have led to an aging population in rural areas. The proportion of the population over 65 (Table 1.1) has increased more rapidly in rural areas than in urban areas. The shortage of access to public services (particularly health services), stated above, causes a real problem for these vulnerable populations.

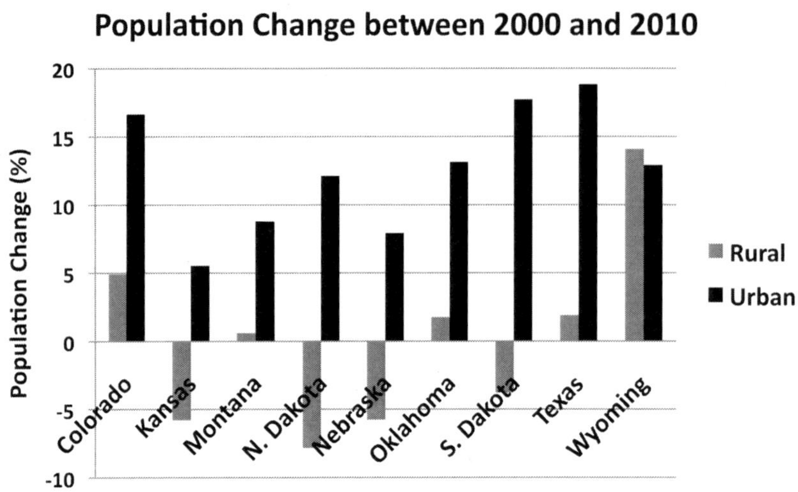

Figure 1.5. Rural-Urban population change rate (U.S. Census Bureau 2010b)

Table 1.1 Percent Population Over 65

	Rural	Urban	Total
Colorado	15.65%	10.94%	14.40%
Kansas	18.87%	14.07%	18.09%
Montana	18.51%	14.98%	18.26%
N Dakota	20.86%	12.02%	20.19%
Nebraska	19.99%	13.14%	19.33%
Oklahoma	16.77%	13.46%	16.04%
S Dakota	18.05%	13.93%	17.61%
Texas	16.99%	12.81%	15.72%
Wyoming	14.60%	12.49%	14.42%

Source: (U.S. Census Bureau, 2010)

Natural Resources

LAND USE AND ECOSYSTEM CONSIDERATIONS

The composition and productivity of native rangelands of the Great Plains are highly dependent upon rainfall and temperature, and range from shortgrass steppe in the west to tallgrass pastures in the east. Large numbers of ungulates co-evolved in the formerly extensive grasslands, with large herds of bison, elk, and pronghorn observed across the Great Plains landscapes in the 1800s. The network of rivers, playas, and wetlands intersecting the Great Plains also provided critical habitats for migratory and wetland bird species. Shortgrass steppe occupies about 108,000 square miles (280,000 km^2) in the Great Plains, from western Texas to the Colorado-Wyoming border (Lauenroth and Milchunas 1991). Most native pastures/rangelands are inappropriate for cropland agriculture because of uneven terrain, poor soil quality, high erosion potential, and/or low rainfall, and many are inaccessible for mechanical harvesting of forage.

The north-south temperature gradients and east-west precipitation gradients also influence the types of dominant agricultural products and production practices implemented across the region. Annual cropping is more dominant in the cooler north and wetter eastern portions of the region. Rangeland and cattle production, while important across the region, is more dominant in the drier western and warmer southern parts of the region. Climate and topography of the Great Plains reduce possibilities for diversification of agricultural practices and land management in any given part of the region. Particularly in the northern part of the region, the short frost-free period reduces flexibility in the number of suitable crops. In the southern and western parts of the region, the cropping season is constrained by precipitation. Rain-fed cropping across the region requires the use of stored soil water because during at least some portion of the growing season precipitation deficits lasting weeks to years occur.

Since the time of settlement, the variable and semi-arid climate has challenged people trying to live off the land. During the 20th century, marginal areas have been ranched or farmed during wet periods, only to be abandoned when dry conditions return. Narrowing profit margins and technology changes have also been driving forces behind the recent trend in farm consolidation in the Great Plains (Lackett and Galvin 2008). The total market value of agricultural products sold in the region is over $92 billion, with 43% of this value coming from crops and 46% from livestock (USDA Economic Research Service 2012b). Although 90% of the land in the region is used for agriculture, the contribution of agriculture to the gross regional product is very small, accounting for roughly two percent (USDA Economic Research Service 2012b).

Currently, the grazing industry in the Great Plains is commercially-oriented and not based on subsistence. The size of ranches is often quite large, and when cattle are pulled off the range they are often finished at feedlots where they are corn-fed. Although, many smaller ranches still exist in the Great Plains, with a number of smaller cow-calf operations being found in the southern portion of the region. Many ranchers also grow crops -- some for sale and some for feed -- so separating livestock operations from cropping operations is often difficult in the Great Plains.

Great Plains land managers are worried about a variety of factors related to climate variability and change, though climate change is often not the most pressing concern (Ojima and Lackett 2002, Lackett and Galvin 2008). More likely, factors such as market or commodity prices, incentives, conservation policy, and social issues are considered in the decision making. In fact, many operators in the region are socio-economically vulnerable due to the declining social services and economic returns for farming and ranching enterprises. Many land use managers and households are operating on the economic margin, and small shifts in climate or markets may drive them out of business. A number of operations have diversified their income streams to provide an economic and household buffer to maintain their ranching or farming enterprise (Hoppe and Banker 2010, Park et al. 2011, Pender et al. 2012). Foreclosures in the region have led agricultural operations to increasingly consolidate into larger enterprises. This leads to population declines in the region, and contributes to the aging of the farm population, as new operators are not coming into the region in great numbers. This trend has put pressure on rural areas, leading to a stressed system where towns may have problems providing adequate social services for inhabitants due to declining population numbers, tax bases, and rural infrastructure.

In the Great Plains environment, weather variability has contributed to the economic dynamics of the cropland and rangeland systems. Regional climate patterns associated with variability in droughts, winter storms, flooding, and other seasonal extreme events have shaped the volatility of agricultural production and social well-being. Reduced summer precipitation and greater proportion of winter precipitation favors deep rooted woody vegetation relative to grass species. These changes in vegetation communities in the foothills and the plains of the region will affect forage availability for domestic and wild grazers. In addition, these conditions also favor certain cool season invasive species, such as cheatgrass (U.S. Climate Change Science Program 2008a). Climate change is already contributing to increased nighttime temperature, increased intensity of rainfall events, extended growing seasons, extended drought periods, and elevated atmospheric CO_2 concentration (Field et al. 2007, U.S. Climate Change Science Program 2008b).

The pace and characteristics of land cover change are highly variable across the Great Plains region (Drummond et al. 2012). This is due in large part to the spatial variability of natural resources and climate patterns, which influence land use management decisions. Land quality (e.g. soil type, topography, erosion), water availability, precipitation and temperature regimes, and other biophysical factors play substantial roles in shaping the broad-scale geographic patterns of crop production, livestock grazing, and other uses. Areas with good soil and favorable climate have a long history of persistent cultivation, while areas that are unsuitable for crops are primarily used as rangeland or may fluctuate between dryland crops and grazing. However, the spatial and temporal dynamics of land use and land cover change are ultimately decided by a combination of landowner decisions, government policies, economic opportunities, population and demographic trends, technological advances, energy and input costs, and evolving agricultural practices.

Much of the Great Plains land cover change results from the episodic expansion and contraction of cropland. Between 1980's and 2000, agriculture to grassland conversions outpaced all other land cover changes combined, though the pace and direction of these conversions varied over time. Factors that contribute to the overall net trends include urbanization in the more populated sections of the plains, cyclic brush clearance in the southern plains, wetland inundation in the northern plains, and other smaller conversions.

Recent trends show that agricultural land cover had a net expansion during the 1970s, which was later reversed by the 1985 Farm Bill's Conservation Reserve Program (CRP).

The CRP provides an economic incentive to convert marginal and environmentally-sensitive cropland to grassland cover or other natural cover types. Historically, government policy is among the important drivers of expansion and decline. However, land cover changes result from interactions among a mix of drivers. For example, the increase of agricultural land cover in the early 1970s was in response to higher grain prices, policies and price supports that favored cropland expansion, and reasonable land prices and interest rates. Technological changes, including the spread of center pivot irrigation, also allowed cropland to expand in areas of water availability, such as the extension of feed corn production and associated industries to the High Plains (Ogallala) Aquifer area. Conversely, local-scale declines in agriculture occurred as urban areas expanded in response to population growth and migration to cities.

Net agricultural expansion was followed by a period of slow rates of land cover change driven by the contraction of export markets, increased costs associated with farm inputs, and high interest rates (Stam and Dixon 2004). By 1985, policy again had a significant effect on land cover as the CRP enabled the conversion of millions of acres of cropland to grassland cover, including in areas of declining groundwater that overlay the High Plains Aquifer. The initial CRP period (1986-1992) had the largest effect on land cover between 1973 and 2000. Currently, farmers are responding to new economic realities, energy policy designed to promote biofuels production, and other drivers that will continue to change the land cover composition of many areas of the Plains, including some key areas of CRP decline in the north central plains and the southern high plains and irrigation decline in parts of the high plains (USDA 2012).

SOILS IN THE REGION

Soils of the Great Plains region are predominantly deep, rich fertile Mollisols. Mollisols form as a result of long-term accumulations of plant material and are high in organic matter content. They are characterized by a thick, dark surface (A) horizon and a high (>50%) base saturation. The development of the dark surface horizon results from the process of "soil formation" involving (1) penetration of plant roots into the soil profile and their subsequent death, (2) decay of organic material, (3) mixing of organic matter by soil micro-organisms, (4) movement of organic and some inorganic colloids within the soil by water (eluviation and illuviation), and (5) formation of resistant organic residues producing the dark color in the soil. Biological activity is important in Mollisols, as soil fauna such as earthworms, insects, and rodents help break down and incorporate organic matter. Clay content is evenly distributed throughout the A and B soil horizons. The movement of clays from the B to the A horizons occurs by a variety of processes, including a common prairie ant (*Formica cinerea).* Mollisols characteristically support grassland or prairie vegetation in climates that have moderate to pronounced seasonal moisture deficits under a wide range of temperature regimes. The typical topography associated with Mollisols is flat or gently rolling to undulating. The parent material is associated with unconsolidated material resulting from glaciation, aeolian deposits (loess) high in calcium, and/or sedimentary rocks such as sandstone, limestone, and shale (Pieper 2005). Mollisols are characteristic not only of the Great Plains, but also of the steppes of Europe, Asia and South America.

In the more humid portions of the Great Plains, soils are dominated by Alfisols. These soils have developed in higher rainfall environments and have undergone moderate leaching and have subsurface accumulation of clay and ≥35% base saturation. These soils are generally occupied by forests, savannas and open prairies.

WILDLIFE, NATIVE VEGETATION, AND CONSERVATION ISSUES

The Great Plains contain a number of natural areas and conservation areas hosting a diversity of wildlife, grassland and wetland ecosystems, and riparian corridors and environmental gradients. These resources support a unique set of birds, fauna, vegetation and insects. The climate gradients and the variability of weather patterns provide a diversity of habitat conditions in support of iconic species, such as the American bison, Greater Sage Grouse, sandhill cranes, ferrets, coyotes, golden eagles, ducks of many kinds, warm water fish populations, pronghorn, horned lizards, amphibians, and others. Climate change and land use patterns across the Great Plains have affected a number of environmental factors and ecosystem services. State and federal wildlife and conservation planners have been developing modified management plans to better incorporate climate issues into their management strategies (The Heinz Center 2008, 2009, Mawdsley 2011).

Native vegetation communities are strongly linked to the gradients of temperature (north to south) and precipitation (west to east) within the Great Plains. Cool-season grasslands in the north give way to warm-season grasslands in the central and southern parts of the region, which in turn transition to drought-adapted shrubs in the southwestern parts and trees in the southeastern parts. As precipitation increases from west to east across the Great Plains, the native vegetation includes more mixed-grass and tall-grass species, and finally a greater number of tree species. Though dominated by grasslands, the Great Plains is also home to a diversity of plants and animals in shrubland, wetland, and forest communities.

At the time of European settlement of the southern Great Plains, woody plants, including eastern red cedar (*Juniperus virginiana*), Ashe juniper (*J. ashei*), Pinchot or redberry juniper (*J. pinchotii*), Rocky Mountain juniper (*J. scopulorum*), and honey mesquite (*Prosopis glandulosa*), were restricted primarily to riparian or deeply dissected areas that seldom experienced fire. However, beginning in the early 20th century, woody plant encroachment into traditional grassland areas has become a substantial land-management issue, that continues to occur at a rapid rate today.

Many plant and animal species have coped with changing climates throughout their evolutionary histories (Axelrod 1985, Elias 1991). Grassland birds, which have persisted through millennia of both climate stasis and extreme variability, date to the early Pliocene, 4.3-4.8 million years ago (mya) (Emslie 2007) when extensive prairie and steppe habitat dominated the Great Plains and climate was relatively stable. With the advent of the glacial-interglacial cycles of the Pleistocene, beginning ca 2.5 mya, the prairie-steppe habitat periodically appeared and disappeared (Emslie 2007). During moister glacial times (such as the late Wisconsin Glacial Period, 15,000-12,000 years before present), areas now covered with grassland were mostly covered by glacial ice or open forests and woodlands with scattered grasslands, as indicated by high levels of tree pollen immediately below the surface (Axelrod 1985). Through the combined impacts of a drier climate, fire, and grazing by large herbivores, the area reverted to extensive grassland interrupted by narrow riparian woodlands along many lakes, creeks, and rivers. During the past 10,000 years -- the Holocene -- relatively moist conditions across grassland landscapes were repeatedly interrupted with droughts intense enough to impact vegetation composition and mobilize sand dunes (Forman et al. 2001).

The rich grasslands of the region have been the basis of a large grazing system for thousands of years and currently support a diversity of native ungulates and other mammals, as well as a diversity of arthropods, reptiles, amphibians, and birds. Although more than 1,100 species of vertebrates have been recorded on the Great Plains, 97 are considered endemic (unique) to the Great Plains, or as having a strong affinity to the plains (Knopf and Smson 1997). Predominant mammals include 16 endemic species, such as the bison (*Bison bison*), pronghorn (*Antilocapra americana*), swift fox (*Vulpes velox*), black-footed ferret (*Mustela nigripes*) black-tailed prairie dog (*Cynomys ludovicianus*) and many other rodent species, and white-tailed jackrabbit (*Lepus townsendii*). Many groups of birds breed across the Great Plains, primarily hawks, grouse, waterfowl, shorebirds, and songbirds. In addition, hundreds of migrant bird species cross the interior during migration from northern breeding areas and southern wintering grounds. Breeding birds endemic to the grasslands include ferruginous hawks (*Buteo regalis*), mountain plovers (*Charadrius montanus*), long-billed curlews (*Numenius americanus*), lark buntings (*Calamospiza melanocorys*), and others.

The wetland basins in the Prairie Pothole region of the Northern Great Plains and the playa lakes region of the Central and Southern Great Plains provide important breeding and migratory habitats for a diversity of wetland-dependent species. Several species of waterfowl nest in grasslands associated with the prairie potholes, notably mallards (*Anas platyrhynchos*), gadwalls (*A. strepera*), and pintails (*A. acuta*). Many species of breeding and migrating shorebirds and other wetland-dependent birds also range across the entire plains region. Prairie wetlands host a multitude of northbound shorebird migrants in spring, the most numerous being the calidridine species, such as semi-palmated sandpipers (*Calidris pusilla*) and white-rumped sandpipers (*C. fuscicollis*) (Skagen et al. 2008).

Amphibians, such as plains spadefoot toads (*Spea bombifrons*) and plains leopard frog (*Rana blairi*), are endemic to the Great Plains. In addition, there are six endemic reptilian species. The amphibians and reptiles of the Great Plains comprise about 20% of the species native to the United States and Canada and feature a mixture of species with primarily southeastern or southwestern distributions and only 10 to 15 endemic species (Corn and Peterson 1996). Reptiles and amphibians rely on their ambient environment to maintain optimal operating temperatures and are sensitive to changes in climate. This is evidenced by a gradient of decreasing species diversity running from south to north and east to west in the Great Plains. Thus, these species are thought to be particularly susceptible to changes in climate (Gibbons et al. 2000), and there is some evidence for climatological impacts to lizards elsewhere (Sinervo et al. 2010). Most of the species diversity is associated with non-grassland habitats, such as permanent water or riparian woodland. However, species of western spadefoots (genus *Spea*) and the Great Plains toad (*Anaxyrus cognatus*) require ephemeral rainwater-filled wetlands for breeding habitat.

Fish habitats include large streams with erratically variable flow, prairie ponds, marshes and small streams, and residual pools of highly intermittent streams. Seven families and 34 species of fish are endemic to the plains, including pallid sturgeon (*Scaphirhynchus albus*), several species of minnow and shiners (family Cyprinidae), madtoms (family Ictaluridae), and darters (family Percidae).

The largest and most diverse class of animals in the Great Plains is insects, including 92 species of dragonflies and damselflies, 220 species of butterflies, and 82 species of grasshoppers that occur in the ecoregion (Ostlie et al. 1997). The many taxonomic groups of aquatic invertebrates and zooplankton include amphipods, copepods, and cladocerans (Wissel et al. 2011).

Many taxa of biota, including plants, insects, and birds, have evolved the capacity to adapt to gradual environmental changes associated with climate, primarily through movement to more favorable areas. An exception to this is the mass extinction of large terrestrial mammals (North American megafauna, including mammoths, mastodons, ground sloths, horses, camels, and others) during the late Pleistocene (ca 11,500-10,500 bp). Although heavily debated, a primary hypothesis for the cause of these abrupt extinctions is the combination of human predation (the arrival of Clovis hunters to North America) coincident with major climatic and environmental changes that had already reduced population sizes (Benedict et al. 1996, Stuart 2008).

WATER RESOURCES

By virtue of its scarcity, water is a critical resource in the Great Plains. Although the region is characteristically dry, humans have managed to transform the land to overcome this limitation. Since water has been a central component of that transformation, a continuous, sufficient water supply is a major concern to inhabitants. Water supply sources include surface water in rivers, streams and lakes, which comes primarily from snowmelt, shallow and deep aquifers, and rain. Drought has always been a factor in the region, with the degree and timing controlled by temperature, precipitation, and the ratio of precipitation to potential evapotranspiration (PET) (Parton et al. 1994). Barry (1983) argues that drought is the key climatic parameter of the Great Plains, as it determines the carrying capacity of the region. Water users in the Great Plains are concerned about a variety of factors related to climate variability and change. Climate change is not the most important concern in this region now, however, as there are many other stresses, including market-driven factors, policy factors, and social factors. In fact, many water users in this region are vulnerable due to the declining reward scale for farming and ranching.

The early 2000's drought was a severe to exceptional event throughout much of the region, and 2002 was the worst drought year on record since 1895 for much of the western Great Plains and the United States (Tronstad and Feuz 2002, Pielke and Roger 2005). During the 2002 drought, a good portion of the central and northern Plains suffered significant agricultural losses: Colorado, Kansas, Montana, Nebraska and South Dakota combined to report an approximate $7.5 billion loss (Knutson et al. 2008). The drought of the early 2000's showed that ranchers, among others throughout the region,

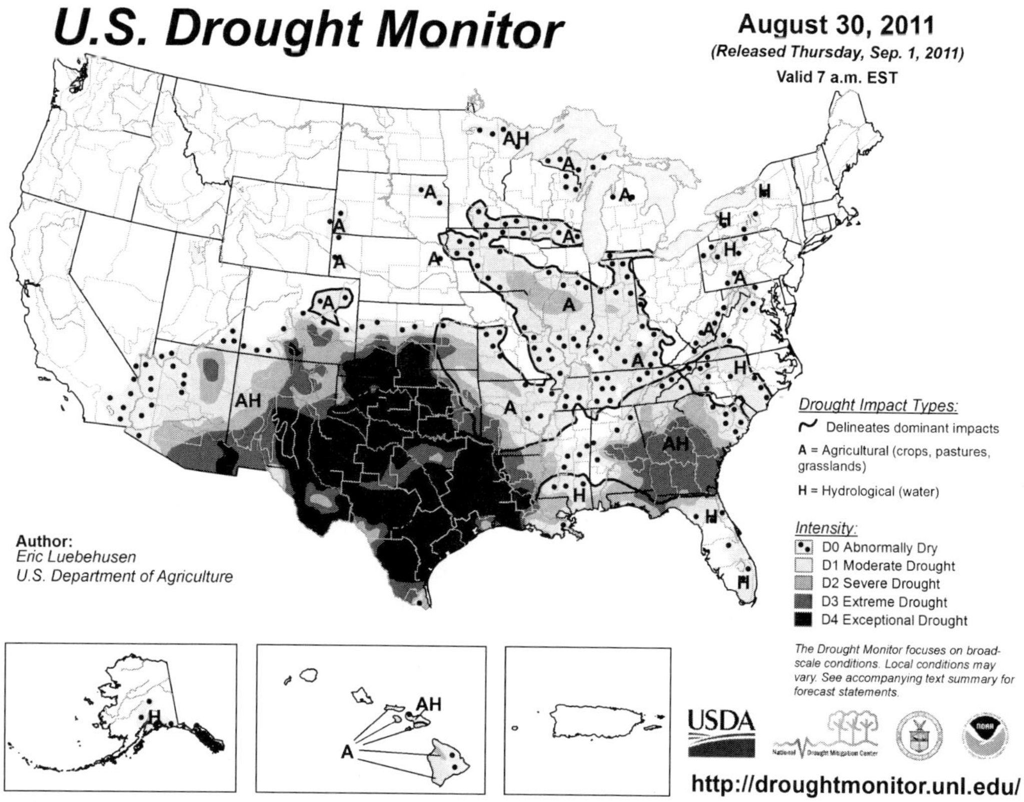

Figure 1.6. The extreme levels of the 2011 drought experienced in the Southern Great Plains indicated a severe and exceptional condition. Some relief in 2012 has occurred due to rainfall into the region.

were dealing with multiple stresses (Nagler et al. 2007), and were largely unprepared for the impacts of prolonged drought (Miller 2005).

The longest recorded drought occurred in the 1950s, and the most disastrous was during the 1930s "Dust Bowl" era. More recently, in 2011 and 2012, the most severe drought in the observational record occurred in the Southern Plains. Texas was the hardest hit state overall with record heat, drought, and fires wreaking havoc over the Southern Plains, and the economy approaching around $10 billion in losses to crops, livestock and timber (National Oceanic and Atmospheric Administration 2011a). In July and August 2011, most of the state of Texas was in "extreme" to "exceptional" drought (Figure 1.6). Using tree-ring records to put this drought in long-term historical perspective (back to 1550), 2011 was only matched in extremity by the year 1789 (National Oceanic and Atmospheric Administration 2011b). However, several prolonged droughts occurred that were similar to the 1950s drought, so while prolonged droughts are even less rare, 2011 is a relatively rare event (National Oceanic and Atmospheric Administration 2011b).

While the 2011 drought is consistent with projections for more intense drought events associated with climate change (IPCC 2012), it still has yet to be confidently determined if it can be attributed to anthropogenic climate change. Experts say that the drought appears to be more strongly associated with natural variability, and specifically the La Niña phase of the El Niño-Southern Oscillation conditions in the Pacific Ocean, as well as the Pacific Decadal Oscillation and the Atlantic Multi-decadal Oscillation. However, it is extremely difficult to definitively understand the role of natural variability versus climate change specific to any one event (pers. comm. Klaus Wolter).

Beyond the major impacts of drought and less moisture, the combination of a lack of water with changes in land use and land cover from agriculture and development practices can lead to deleterious effects (Cook et al. 2009). Local responses will depend on the household characteristics and the availability of public assistance associated with local to regional policy mechanisms in place to enhance coping mechanisms and to reduce vulnerability (Kallis 2008). Groundwater depth also plays an important role in the regional effects of drought, since precipitation minus evaporation anomalies show a strong dependence on convergent flows and water-table depth (Maxwell and Kollet 2008).

In Nebraska, research shows that the most vulnerable areas to agricultural drought were non-irrigated cropland and rangeland on sandy soils, located in areas prone to season water deficits (Wilhelmi and Wilhite 2002). The identification of drought vulnerability is critical to development of appropriate preparedness options and mitigation-oriented drought management strategies (Wilhite and Pulwarty 2005, Wilhite et al. 2007). Research on Nebraska farmers involved in sustainable agriculture organizations reported a range of implemented practices to reduce their drought vulnerability, such as organic soil building techniques, reduced tillage, targeted crop selection, and diversification of crop and livestock production systems (Knutson and Haigh 2011). Those same farmers reported a number of barriers to adapting to drought risk, such as a lack of capital and market variability and responses (Knutson and Haigh 2011). Incorporating these non-climatic variables into science and policy responses to potential increased drought from climate change will be vital to farmers' viability in the Great Plains region and throughout the United States.

Precipitation gradients are also very strong across the region with mean annual deposition ranging from 12 in (30 cm) in the short-grass steppe along the foothills of the Rocky Mountains, to more than 39 in (100 cm) per year approaching the Mississippi River. Precipitation is projected to increase in the north and decrease in the southern high plains, including potential shifts in snowpack, spring rainfall and extreme events. Water availability and droughts here can critically affect threatened regional water resources, including the Ogallala (High Plains) Aquifer, which are essential for agriculture, natural systems, protected species, and the health and prosperity of its citizens.

The aquifer receives recharge from precipitation, which mixes with "ancient" water that has been stored in subterranean basins since it washed down from the Rocky Mountains during the last ice age. Rainfall is not always sufficient, even with existing surface water impoundment facilities, to support the demand necessary to maintain present agricultural yields, particularly in the western portion of the Great Plains (Norwood 2000). Considerable supplementation has been provided through irrigation from aquifers. This makes aquifer depletion a serious concern in some areas of the region because their depletion rate is often faster than the rate of recharge (McMahon et al. 2007, McGuire 2011).

As population increased in the Great Plains and irrigation became widespread during the past 60 years, annual water withdrawals began to outpace natural recharge (McGuire 2007). Approximately 19 billion gallons (72 billion liters) of groundwater are pumped from the aquifer daily to irrigate 13 million acres of land and provide drinking water to more than 80 percent of the High Plain's population (Dennehy 2000). Since 1950, aquifer water levels have dropped an average of 13 feet (4 m), equivalent to a 9 percent decrease in aquifer storage. In heavily irrigated parts of Texas, Oklahoma, and Kansas, reductions are much greater, from 100 feet (30 m) to over 250 feet (76 m). Projections of increasing temperatures, faster evaporation rates, and more sustained droughts brought on by climate change will only add more stress to overtaxed water sources (Green et al. 2007, Gurdak et al. 2007, Lettenmaier et al. 2008, U.S. Climate Change Science Program 2008a). Current water use on the Great Plains remains unsustainable, as the High Plains Aquifer continues to be tapped faster than the rate of recharge. Without the irrigation buffer of the aquifer, agriculture on the High Plains may become tenuous, and land-use changes, including abandonment of formerly productive croplands, may be induced by lack of water availability. It is unclear, at this time, what role these lands could have in the adaptive response of Great Plains ecosystems to climate changes.

In the Great Plains as a whole, crop and pasture land contributes 49% (2024 billion ft^3 per year or 57.3 billion m^3 per year) of the water supply (compared to a national average of 26%), followed by rangeland (911 Bft^3 per year or 25.8 Bm^3 per year), forest (703 Bft^3 per year or 19.9 Bm^3 per year), and wetlands (314 Bft^3 per year or 8.9 Bm^3 per year) (Brown et al. 2008). Seasonality is an important factor affecting water systems in the Great Plains. The flow of these waters has been altered by humans through diversion, impoundment, and irrigation for urban and agricultural uses.

Precipitation in the Northern Great Plains is projected to increase from climate change, leading to more flooding events in some areas (U.S. Bureau of Reclamation 2011). Heavy precipitation could increase as much as 30% in South Dakota, which has already seen considerable flood damage recently with nine flood disaster declarations the past decade (FEMA 2012). Wu et al. (2012) used the Soil and Water Assessment Tool model to assess the effects of elevated atmospheric CO_2 concentrations on historical and projected hydrological changes in the Upper Mississippi River Basin, and found that approximately 1–4% of the streamflow in the Upper Mississippi River Basin during 1986 through 2008 could be attributed to the elevated CO_2 concentration. The same study also projected increased spring water yield and soil moisture and a substantial decreased summer water yield and soil moisture for 2071 to 2100, which could lead to both increased flooding and droughts (Wu et al. 2012). However, it is important to note that even without climate change scenarios more persistent flooding and drought periods were common in the Great Plains before the 1800s as determined by studies of paleo-records from tree-ring and lake-sediment data (Shapley et al. 2005). Decadal climate variability in the Missouri River Basin is also known to have strong tele-connections to oceanic-atmospheric oscillation patterns that affect water yield in some locations (Mehta et al. 2011). In other words, it will be important to understand both climate variability and climate change for the development of early warning systems for variable streamflows and both floods and droughts, as well as planning efforts for future water projects (Knutson et al. 2008).

In addition, point and nonpoint source pollution have introduced a wide array of organic chemicals, toxic metals, and fertilizers, such as nitrogen and phosphorous, into Great Plains aquatic ecosystems. Several factors account for water pollution in the Great Plains, including extraction processes; farm management practices associated with fertilizer usage, pesticide applications, manure and sediment run-off; industrial run-off; and inflow from built environment. The pollution leads to increased salinity, nutrient loading, turbidity, and siltation of streams. Shallow aquifers also suffer from these pollution problems (USDA Natural Resources Conservation Service 1996). Drinking water quality is reduced as a result of pollution, particularly in rural communities, where the water supply is taken from local sources and not from municipal treatment systems. These water supplies are more vulnerable to runoff and leaching of agricultural chemicals. This decrease in water quality has affected food production, human drinking water supplies, and wildlife habitat. Alteration of vegetation, introduction of nonnative plant and animal species, and over-harvesting of native species has also damaged these aquatic ecosystems.

MAJOR RIVER BASINS

The Great Plains are transected by four major river systems: the Red River of the North in the northeast, the Missouri River in the north and central region, the Arkansas-Red river system draining the central region, and the Texas Gulf Basin, including the Rio Grande River, in the south. These river systems have served as passageways into and across the Great Plains. They continue to serve as critical resources for energy, irrigation, and conservation efforts throughout the Plains region. An overview of these major river systems is provided here.

Red River of the North

The Red River of the North originates along the North Dakota-Minnesota border. The river flows in a northward direction for 545 miles (877 km) through the Red River Valley, containing cities such as Fargo-Moorhead and Greater Grand Forks, before eventually entering Manitoba where it discharges into Lake Winnipeg and ultimately into Hudson Bay (Benke and Cushing 2005). The Red River's 48,490 square-mile (125,589 km^2) drainage area, which includes the Devils Lake sub-basin, is near the geographic center of North America and includes portions of North Dakota, South Dakota, and Minnesota as well as parts of Manitoba and Saskatchewan (Benke and Cushing 2005). The Red River Valley is part of what used to be the extremely flat floor of ancient glacial Lake Agassiz, and the river has a remarkably low gradient that can be as little as 1.5 inches per mile (2.4 cm/km) in some reaches.

The flat topography of the Red River Valley combined with the synchrony of a northward flowing river and a northward moving spring thaw makes this region one of the most flood-prone areas in the US. Runoff from the warmer southern portion of the Valley progressively joins with fresh, melted waters from more northerly reaches. These flows may then get dammed by natural ice jams downstream. River water then overflows, spreading across and flooding the flat former lakebed. Based on more than

100-year-old river stage data collected in Fargo, the Red River exceeded the major flood stage -- the point at which extensive inundation of structures and roads is expected to occur -- 16 times over this 100 year old record.

Missouri River

In the northern and central Great Plains, the dominant river system is the Missouri. Originating in the northern Rocky Mountains of southwestern Montana near the city of Three Forks and has contributions from the Platte River draining the Colorado and Wyoming. The river flows over 2,300 miles (3,701 km) in a southeasterly direction through Montana, North Dakota, South Dakota, Colorado, Nebraska, Iowa, Kansas, and Missouri, finally discharging into the Mississippi River near St. Louis (Benke and Cushing 2005, Reclamation 2011). The Missouri is typically identified as the longest river in the US and the longest named river in North America (Benke and Cushing 2005). Thirty-seven tributaries flow into the Missouri including the Yellowstone, White, Platte, and Gasconade rivers(Benke and Cushing 2005). In addition to part of two Canadian provinces, the river drains over 500,000 square miles (1,295,000 km^2) consisting of all or part of ten states and 25 Native tribal reservations or lands. Its basin comprises roughly one sixth of the land area of the lower 48 states (Benke and Cushing 2005, U.S. Bureau of Reclamation 2011).

The drainage area of the river consists mainly of two physiographic divisions that vary greatly in terms of climate. One is the Rocky Mountain system where total annual precipitation in the mountains averages over 31 inches (80 cm) and often falls as snow (Benke and Cushing 2005). The largest portion of the Missouri River watershed, though, falls within the semiarid Great Plains where total annual precipitation averages just 14 inches (36 cm) (Benke and Cushing 2005). Thus, despite its length and large watershed, the Missouri's average discharge at its mouth is less than the discharges of other rivers such as the Ohio and Columbia (U.S. Geological Survey 1990). In addition to water sources rising within the basin, water is also transferred from the Colorado River to Northern Colorado via the Colorado-Big Thompson and Windy Gap projects. The water transfer is for agricultural, industrial, municipal, and hydroelectric power purposes (Northern Water 2012a, 2012b, U.S. Bureau of Reclamation 2012a, 2012b).

Discharge patterns in the Missouri main-stem reflect the influence of both the Rocky Mountains and the Great Plains physiographic divisions. Main-stem flows start to rise in March with the melting of prairie snow and then peaks in June due to a combination of Rocky Mountain snowmelt and late spring precipitation on the Plains (Benke and Cushing 2005, U.S. Army Corps of Engineers 2006). Discharge then declines in July. Although system regulation helps reduce flooding, if floods do occur, they typically occur between March and July (U.S. Army Corps of Engineers 2006). Portions of the Missouri Basin have experienced massive flooding events associated with unusual weather patterns, contributing to heavy rainfall concurrent with rapid snowmelt during spring 2011 (Knutson et al. 2008, National Oceanic and Atmospheric Administration 2012).

Arkansas-Red River System

The Arkansas-Red River system consists of the Arkansas River and the Red River of the South, which are the two main rivers draining the Central and Southern Great Plains region. The Arkansas River flows from the Rocky Mountains of central Colorado, near the city of Leadville and some of the tallest peaks in the lower 48 states (Benke and Cushing 2005). It flows approximately 1,460 miles (2,350 km) in a generally east/southeasterly direction through the Royal Gorge in Colorado, the states of Kansas and Oklahoma, and into Arkansas where it discharges into the Mississippi River near the town of Napoleon (Benke and Cushing 2005, Statewide Water Quality Management Plan 2011). It drains an area of roughly 161,000 square miles (416,988 km^2) (Benke and Cushing 2005). In addition to water sources rising within the basin, the Arkansas River also receives snowmelt runoff imported from Colorado's West Slope across the Continental Divide to the state's semi-arid east slope, via the conduits, tunnels, and reservoirs of the Bureau of Reclamation's Fryingpan-Arkansas Project, completed in 1990, as well as through several other non-federal diversion projects (Muller and Smith 2000, U.S. Bureau of Reclamation 2010).

The source waters of the Red River of the South arise among streams flowing through the Texas Panhandle (Benke and Cushing 2005). As the river travels east towards Wichita Falls, Texas, it drains some of the driest regions in the Southern Plains, which receive less than 20 inches (51 cm) of rainfall per year. As a result, the river may experience extended "no flow" periods and pooling up (Benke and Cushing 2005). The Red River of the South becomes more substantial as it continues eastward past Wichita Falls and enters Lake Texoma, a reservoir shared by Oklahoma and Texas and formed by the Denison Dam (Benke and Cushing 2005). The Red River forms the long-debated boundary between these two states. The river exits Lake Texoma and continues traveling east/southeast towards Louisiana where it ends at the confluence of the Old and Atchafalaya Rivers, the latter of which empties into the Gulf of Mexico (U.S. Army Corps of Engineers 2002). The total length and drainage area reported for the Red River of the South vary somewhat. However, a U.S. Geological Survey Fact Sheet (1990) lists the river's length as 1,290 miles (2,076 km) and its watershed as 93,200 mi^2 (241,387 km^2).

The Army Corps of Engineers has undertaken a Red River Basin Chloride Control Project to reduce naturally occurring brine fluxes in several Texas and Oklahoma subbasins, the goal being to improve water quality for municipal, industrial, and agricultural uses (U.S. Army Corps of Engineers 2010). Although there would be benefits from the water quality viewpoint, water withdrawals from the river would possibly increase the number of no-flow days in the upper basin (Benke and Cushing 2005). In addition, changes in the river's natural salinity regime could affect river ecology (Benke and Cushing 2005).

Texas Gulf Basin

In the Texas Gulf Basin, eleven major rivers traverse through Texas and discharge into the western Gulf of Mexico (Benke and Cushing 2005). Two of these are the Rio Grande and Trinity Rivers (RONA). The Rio Grande rises in the San Juan Mountains of Colorado (part of the Rocky Mountains) and flows south through New Mexico, passing Albuquerque on its way towards El Paso/Ciudad Juarez (Benke and Cushing 2005). From there, it flows generally southeast forming the international boundary between Texas and Mexico as it travels towards the Gulf of Mexico where it discharges near Brownsville, Texas

(U.S. Bureau of Reclamation 2011). Along the way, it passes through several reservoirs, including the Cochiti, Elephant Butte, Caballo, Amistad, and Falcon. Two stretches of the river have been declared part of the National Wild and Scenic Rivers system, including one reach running through Big Bend National Park (Benke and Cushing 2005). The total length and drainage area reported for the Rio Grande vary somewhat. However, according to a USGS fact sheet (U.S. Geological Survey 1990), the river has a length of 1,900 miles (3,050 km) and a combined US-Mexico drainage area of 336,000 square miles (870,000 km^2).

The mountain headwaters region of the Rio Grande receives about 40 inches (102 cm) of precipitation per year, mostly as snow (U.S. Bureau of Reclamation 2011). Snowmelt is the main source of water for the river and dominates the hydrograph for the upper portion of the Rio Grande, with peak flows typically occurring in spring and early summer (Benke and Cushing 2005, U.S. Bureau of Reclamation 2011). However, as the Rio Grande passes through multiple reservoirs further downstream, the reservoirs become the controlling factor in the river's hydrograph (U.S. Bureau of Reclamation 2011). In recent years, increased human consumption of Rio Grande water by both the US and Mexico has resulted in intermittent or lower flows reaching the downstream sections, and, in 2002 and 2003, Rio Grande waters did not reach the Gulf of Mexico for multiple months (Benke and Cushing 2005, U.S. Bureau of Reclamation 2011). Irrigation is a major water demand. Important issues in the Rio Grande basin include endangered species and water quality issues, such as salinity (U.S. Bureau of Reclamation 2011). The Rio Grande Compact Commission is undertaking a multi-state salinity control program, modeled after the Colorado River Salinity Control Forum (D. and Lewis 2008).

The 715 mile (1,151 km) -long Trinity River starts in the Four Forks region in the north-central/northeastern part of Texas. The Clear Fork and West Fork of the river join near Fort Worth, the Elm Fork near Dallas, and the East Fork just south of Dallas (Benke and Cushing 2005, Trinity River Authority of Texas 2010). The river then flows generally southeast where it discharges into the Gulf of Mexico. The Trinity River provides water for two of the most populous metropolitan areas in the US (Houston and Dallas/Fort Worth), and the river empties into Galveston Bay, one of the nation's most productive ecosystems and commercial fisheries (Trinity River Authority of Texas 2010, Mackun and Wilson 2011).

The Trinity River drains 18,000 square miles (46,600 km^2) and is the largest river basin in Texas that lies entirely within the state. Most of the flow in the river comes from rainfall runoff. Precipitation varies within the basin ranging from 29 inches (74 cm) per year in the west to 53 inches (135 cm) per year closer to the coast (Trinity River Authority of Texas 2010). Flows in basin streams are quite variable and can be very low during the summer. In order to provide a more stable water supply, a total of 31 reservoirs have been built on the river and its tributaries (Trinity River Authority of Texas 2010). In addition, seven reservoirs outside the watershed either provide water to Trinity basin users, or are under contract to do so in the future (Trinity River Authority of Texas 2010). Because of groundwater scarcity, Trinity basin users must rely on surface water (Trinity River Authority of Texas 2010).

Chapter 2

Characteristics of Agricultural System and Energy Resources

Agricultural System

The Great Plains produces much of the nation's food and fiber. The region produces nearly two-thirds of the nation's wheat, more than half its beef, a fifth of its corn, a quarter of its cotton, four-fifths of its grain sorghum, and a sixth of its pork (Duncan et al. 1995). While wheat and beef production are important across most or all of the Great Plains states, one or more of the states also contribute significantly to production of other animal (hogs, dairy, broilers - i.e. chickens raised for meat, and sheep) and crop (corn, soybean, cotton, sorghum, canola and other) commodities (Table 2.1). Changes in land use management, climate, and hydrological extremes will impact how natural resources will be utilized and sustained over time in the Great Plains, affecting the region's social wellbeing and ecosystem integrity.

In the nine Great Plains states (Colorado, Kansas, Montana, Nebraska, North Dakota, Oklahoma, South Dakota, Texas, and Wyoming) there are approximately 510,405 farms and 340,653,196 total acres (1,378,575 km^2) in farms (USDA National Agricultural Statistics Service 2009). Approximately 42% or 143 million acres (578,700 km^2) is in cropland and approximately 52% or 178 million acres (720,300 km^2) is in permanent / native pastures. Of the 143 million acres (578,700 km^2) of cropland, in 2007 over 22 million acres (89,000 km^2) were planted to corn, over 4.8 million acres (19,400 km^2) were planted to cotton, over 5.6 million acres (22,700 km^2) were planted to sorghum, over 14.2 million acres (57,500 km^2) were planted to soybeans, and over 29.5 million acres (119,400 km^2) were planted to wheat (USDA National Agricultural Statistics Service 2009). An additional 14.8 million acres (59,900 km^2) of cropland were in improved pastures and 15 million acres (60,700 km^2) of Great Plains farmland were in the CRP program.

BEEF CATTLE PRODUCTION

Because of the vast quantities of native rangelands, livestock production (mostly beef cattle) is one of the most important sectors in US Great Plains agriculture, both economically and socially. On average, 30% (North Plains) to 68% (South Plains) of total farm production value in the Great Plains comes from beef cattle (McBride and Matthews 2011).

The total number of ungulates grazing the Great Plains today is estimated to be similar to the numbers before European settlement (Table 2.2). Essentially, wild ungulates have been replaced with domesticated ungulates. In contrast to pre-settlement, livestock

Table 2.1 Value of Top 5 Agricultural Commodities by State

		Value of receipts thousands $	Percent of state total farm receipts	Percent of US value
Colorado	1. Cattle and calves	2,852,521	47.4	5.5
	2. Corn	604,082	10	1.3
	3. Wheat	500,407	8.3	4.6
	4. Dairy products	456,740	7.6	1.5
	5. Hay	287,127	4.8	5.3
Kansas	1. Cattle and calves	6,533,521	46.8	12.7
	2. Corn	2,118,661	15.2	4.7
	3. Wheat	1,724,662	12.4	15.9
	4. Soybeans	1,470,992	10.5	4.4
	5. Sorghum grain	673,287	4.8	50.4
Montana	1. Cattle and calves	1,084,644	35.6	2.1
	2. Wheat	1,032,557	33.9	9.5
	3. Hay	267,970	8.8	5
	4. Barley	157,348	5.2	21.2
	5. Lentils	77,593	2.5	37
Nebraska	1. Cattle and calves	7,193,865	41.6	14
	2. Corn	5,347,448	30.9	11.9
	3. Soybeans	2,647,762	15.3	8
	4. Hogs	815,836	4.7	4.6
	5. Wheat	326,594	1.9	3
North Dakota	1. Wheat	1,901,364	28.7	17.5
	2. Soybeans	1,247,264	18.9	3.8
	3. Cattle and calves	731,092	11.1	1.4
	4. Corn	665,142	10.1	1.5
	5. Canola	356,746	5.4	90.3

Table 2.1 Value of Top 5 Agricultural Commodities by State (cont.)

State	Commodity	Value of receipts thousands $	Percent of state total farm receipts	Percent of US value
Wyoming	1. Cattle and calves	2,984,670	48.5	5.8
	2. Broilers	724,446	11.8	3.1
	3. Hogs	696,411	11.3	3.9
	4. Wheat	533,510	8.7	4.9
	5. Dairy products	171,000	2.8	0.5
Texas	1. Corn	2,065,603	26.9	4.6
	2. Cattle and calves	2,002,387	26	3.9
	3. Soybeans	1,588,307	20.7	4.8
	4. Wheat	657,325	8.6	6
	5. Hogs	455,370	5.9	2.5
South Dakota	1. Cattle and calves	7,564,446	38	14.7
	2. Cotton	2,589,126	13	41.3
	3. Broilers	1,757,613	8.8	7.4
	4. Dairy products	1,505,313	7.6	4.8
	5. Greenhouse/nursery	1,311,139	6.6	8.4
Oklahoma	1. Cattle and calves	732,883	62.5	1.4
	2. Hay	122,520	10.5	2.3
	3. Hogs	71,070	6.1	0.4
	4. Sugar beets	44,252	3.8	2.7
	5. Sheep and lambs	34,604	3	6.5

Source: (USDA Economic Research Service, 2012)

Table 2.2 Estimated populations of wild and domestic ruminants in the Great Plains today and in the 15th century

Species	Pre-European settlement	Current
Bison	30,000,000 to 75,000,000	500,000
Elk (wapiti)	10,000,000	1,000,000
White tailed deer	30,000,000	25,000,000
Mule deer	13,000,000	4,000,000
Beef cattle	0	64,800,000
Dairy cattle	0	13,800,000
Sheep	0	5,700,000
Goats	0	3,100,000
Total	83,000,000 – 128,000,000	117,900,000

Source: (Hristov, 2012)

animals on native pastures are frequently supplemented with mineral, energy and/or protein feeds to improve reproduction and animal growth.

Grazing animals, both domesticated and wild, play a vital role in the ecology of grasslands by providing an efficient means of recycling plant and soil nutrients. Ruminants, such as beef cattle, can consume fibrous feeds and byproducts of other industries, including grain ethanol, soybean oil and cottonseed oil that are unfit for human consumption, and turn them into high-quality foods. Livestock can potentially affect climate change primarily via enteric and manure-based GHG (greenhouse gas) emissions. However, they may also be affected by climate change. Livestock production (especially grazing) systems in different parts of the US and world have evolved over long periods of time to fit local environmental conditions, such as water and forage availability (Reynolds et al. 2010). Retaining livestock grazing systems is important to provide economic returns landowners who retain native grasslands in the landscape.

The US and Great Plains beef cattle industry is comprised of four major sectors: 1) cow-calf, 2) stocker, 3) feedlot, and 4) packer. Approximately 40% of US beef cows, 75% of all US feedlot cattle, and 50% of US domesticated bison are in the Great Plains. The cow-calf, stocker, feedlot, and packer segments of the US cattle industry are inexorably linked, and changes in one sector can have major impacts on the other sectors (Galyean et al. 2011).

In the Great Plains, approximately 18.4% of cow-calf operations have fewer than 50 cows, compared to 24.9% in the bordering states and 28.7% nationally. Over 60% of the operations in the Great Plains have over 100 beef cows (Table 2.3). These cow-calf operations occur primarily on native rangelands because they provide an efficient means of harvesting the available forage. The cow-calf herds are a year-round system that must

Table 2.3 Typical size of beef cow operations: % of operations

Number of cows	Great Plains	Bordering States	US Average
< 50 cows	18.4	24.9	28.7
50-99 cows	16.7	16.6	17.2
100-199 cows	22.0	17.2	17.5
200-499 cows	27.6	21.8	20.5
500 cows or more	15.3	19.4	16.1

Source: (USDA National Agricultural Statistics Service, 2007)

live within nutritional constraints of the ecoregion in order to be economically and ecologically sustainable (McBride and Matthews 2011, Phillips et al. 2011). Supplemental feed is often provided during seasons where forage is lacking and these protein/energy supplements are often comprised of byproducts of the corn milling (distillers grains) or vegetable oil (i.e., soybean meal and cottonseed meal) industries.

Changes in forage availability or quality caused by climate change can alter the supplemental feed strategies needed to maintain animal production. Over 70% of beef calves in the US are born between January and April (Phillips et al. 2011) - typically termed "spring calving." In most of the Great Plains, this is when pastures begin their spring growth and have their highest nutritional value. This provides lactating cows with adequate nutrition to replenish body stores that are lost during the winter and at calving. The spring forage also provides sufficient energy and protein for milk production. About 80% of US beef cows wean a calf each year (McBride and Matthews 2011). On average, each cow-calf unit requires 11 to 13 acres (4.5 to 5.3 hectares) on the Great Plains; these values may range from 30 or more acres (12 or more hectares) in the arid west to 3 to 5 acres (1.2 to 2 hectares) in the east. This compares to 3 acres (1.2 hectares) per cow-calf unit in the North Central and Southeast regions of the US and over 19 acres (7.7 hectares) in the far west.

A small percentage of cattle are finished on pastures, rather than in feedlots. The biggest challenge grass-finished beef producer's face is having a high-quality supply of forage available for 12 consecutive months. On average, cattle in feedlots are fed for approximately 150 days before going to slaughter. They typically consume about 20 pounds (9 kg) of feed dry matter each day, gain 3 to 4 lbs (1.4 to 1.8 kg) of body weight each day, and require approximately 5.0 to 6.5 lbs (2.3 to 2.9 kg) of feed dry matter for each lb of weight gain. Typical feedlot diets today will contain from 20 to 80% corn grain and up to 60% byproducts, such as distillers grains or gluten feed (Vasconcelos and Galyean 2007). Approximately 3,969,400 acres (16,064 km2) (about 5% of US corn acreage) are required to produce the corn used annually by the US cattle feeding industry to feed 22.3 million head (USDA 2011). In 1961, a producer used approximately 0.6 acres (0.24 hectares) of farmland per person to produce enough feed for meat, dairy and poultry consumption; in 2005, that had declined to approximately 0.27 acres (0.11 hectares) (Elam 2007). In 1960, approximately 80% of US grain and soybean acres were used for livestock feed production. By 2005, the amount fed to livestock had declined to 50% due to enhanced crop yields. Today, approximately 2 acres (0.8 hectares) of cropland produce enough feed to produce one ton of meat and poultry production.

CROP PRODUCTION

A number of crops are produced throughout the Great Plains with the distribution of crops varying according to climatic gradients. Plants with the C_3 photosynthetic pathways (see box 2.1) tend to grow better in the cooler, wetter northern and eastern region and C_4 plants thrive in the southern and western region. Irrigation, however, has allowed the expansion of corn and wheat in the west and south (Tieszen et al. 1997). The major harvested crops are wheat (accounting for 50% of harvested land), hay (20%), corn (15%) and cotton (4%) (Parton et al. 2007). Wheat production in the Great Plains is the most productive wheat region in the world. Forty% of the country's sorghum, 36% of its barley, 22% of its cotton, 14% of its oats and 13% of its corn are produced in the region (Ojima et al. 1999).

The most important factors contributing to the increased productivity in the Great Plains include: increased irrigation, pest management and fertilizer application, improved tillage practices, and improved plant varieties (Parton et al. 2007). Tillage, utilized for all crops in the region, is the physical loosening of soil to optimize conditions for germination, seedling establishment and crop growth (Lal 1979). The benefits of tilling include seedbed preparation, weed control, evaporation suppression, water infiltration enhancement, and erosion control (International Board for Soil Research and Management 1990). Increases in irrigation and herbicide use have caused a shift in practices away from traditional tillage. Tillage reduction increases water and energy efficiency, carbon sequestration and nutrient retention.

Technological improvements and yield increases come at a cost. The proportion of farm income spent on agricultural inputs (fertilizer, herbicides, insecticides and energy use) has steadily increased since the 1950s (Parton et al. 2007). Inputs accounted for 30% of gross farm income in 1949 and more than 60 percent by the 1990s. This rise in cost has reduced the potential for profit, despite exponential yield increases. Although input costs have increased, profit predictability has remained stable and risk has been lowered due to better technology, increased irrigation and government payments to farmers, which have increased more than 60% since the 1980s (Parton et al. 2007).

In addition to economic costs of inputs, there are environmental costs. Fertilizer application is commonly used across the region, however over-application of fertilizer (particularly nitrogen) causes leaching of nutrients and eventually eutrophication of waterways (Rabalais et al. 2002). Nitrous oxide (N_2O) emissions, a strong greenhouse gas, and nitrate (NO_3) leaching tends to be lower in the western part of the region and increases toward the wetter eastern portion. This is because both increase in precipitation across the west to east gradient associated with greater crop intensity and fertilizer usage. (Parton et al. 2007). For cropland, a primary greenhouse gas emission of concern is N_2O, nitrous oxide, associated with fertilizer and manure application. Ribaudo et al. (2011) found that nitrogen management in the Northern Great Plains failed to meet

BOX 2.1

C_3 & C_4 Differential Responses to CO_2 & Temperature

The vast majority of plant species in agronomic and grassland ecosystems in the Great Plains belong to two photosynthetic classes of plants, C_3 and C_4. Plants with the C_3 photosynthetic metabolism account for over 95% of Earth's plant species, and include most crop species (e.g. rice, beans and wheat). They are found within diverse environmental conditions, but often perform best under moderate temperature and light conditions and when water is relatively abundant. In contrast, C_4 plants (e.g. corn and sorghum) comprise less than 5% of Earth's plant species, have characteristically high water use efficiency, and thrive under high light and temperature conditions. C_4 grasses are an important component of grasslands and savannas, and C_4 crops produce 40% of the world's grain. Due to differences in photosynthetic pathways, rising CO_2 concentrations are expected to directly enhance photosynthesis and therefore growth of C_3 plants, but have little direct effect on C_4 photosynthesis (Ainsworth and Long 2005). Rising CO_2 also closes the stomatal pores in most plant species, C_3 and C4 alike (Wand et al. 1999), which reduces water loss and improves plant water use efficiency (Morgan et al. 2004, Leakey 2009). Thus, rising CO_2 concentrations have the potential to enhance photosynthesis and growth of C_3 plants, but will likely only enhance growth of C_4 plants under water-limited conditions when high water use efficiency is adaptive. Warming increases plant water loss and stress, but may favor warm-season C_4 plants. The combined effects of rising CO_2 and climate change on plant production and species responses are complex and likely to affect C_3 and C_4 plants differently, depending on present-day conditions (warm versus cool, wet versus dry) and the degree and pace of global changes.

conservation criteria due to rate (28%), timing (15%) or method of application (45%). In the Southern Great Plains failure to meet conservation criteria for rate, timing, or method were 32%, 38%, and 18%, respectively.

In the semiarid portion of the Great Plains, dryland wheat farming has been made possible mainly by fallow systems, in which only a portion of an operator's land is planted each year and the rest is left idle to accumulate water and nutrients for subsequent crops (Ojima and Lackett 2002). Wheat fallow is a common practice in the western Great Plains and provides farmers with a reliable income and stable yields from year to year (Croissant et al. 2008). However, the fallow system has a low water use efficiency and results in declining soil organic matter and increased soil nitrous oxide fluxes to the atmosphere. Reduced tillage systems allow more diverse crop rotations with less frequent fallow, which leads to increased precipitation-use efficiency and enhanced soil function (Westfall et al. 2010). In long-term cropping studies in eastern Colorado, annual grain production from no-till systems with less frequent fallow improved by 75%, and economic return increased by 13% to 36%, compared with the traditional wheat-fallow cropping system.

Irrigated cropping is important in all of the Great Plains states. In some states, it is based on groundwater pumping, dominated by irrigation from the Ogallala Aquifer

Table 2.4 Freshwater Withdrawals in Great Plains States

	ND	SD	NE	KS	OK	TX	MT	WY	CO
Irrigation, MGD (m^3 per day)	151 (572)	292 (1105)	8460 (32025)	2740 (10372)	495 (1874)	7800 (29526)	9670 (36605)	3990 (15104)	12300 (46561)
Livestock, MGD (m^3 per day) [1]	23 (87)	48 (182)	108 (409)	108 (409)	162 (613)	258 (977)	39 (148)	16 (61)	33 (125)
Public Supply, MGD (m3 per day)[1]	67 (254)	100 (379)	330 (1249)	403 (1526)	646 (2445)	4270 (16164)	142 (538)	96 (363)	864 (3271)
Domestic, MGD (m^3 per day) [1]	9 (34)	8 (30)	52 (197)	15 (57)	25 (95)	257 (973)	24 (91)	6 (23)	29 (110)
Surface water for irrigation, MGD, (m^3 per day)	73 (276)	143 (541)	1150 (4353)	114 (432)	134 (507)	1680 (6359)	9530 (36075)	3570 (13514)	10000 (37854)
Groundwater for irrigation, MGD, (m^3 per day)	78 (295)	149 (564)	7310 (27671)	2620 (9918)	361 (1367)	6120 (23167)	140 (530)	422 (1597)	2320 (8782)

1 – Combined surface water and groundwater withdrawals
Source: (Kenny et al., 2009)

which extends from the Texas High Plains through the Oklahoma Panhandle, western Kansas, eastern Colorado, Nebraska, into southern South Dakota. In other states, there is a significant amount of irrigation from surface water supplies, primarily from major water projects managed by the Bureau of Reclamation (Table 2.4).

Most Great Plains cropland has undergone loss of soil carbon compared to uncultivated prairie soils (Haas et al. 1957, Hartman et al. 2011). Adoption of no-tillage systems and increased crop intensity may have the potential to enhance soil carbon sequestration. Many studies have shown that soil carbon is increased in surface soil layers in no-tillage systems, compared to conventionally tilled systems across the Great Plains (Potter et al. 1997, G.A. et al. 1998, Sainju et al. 2006, 2011, Blanco-Canqui et al. 2011). However, Blanco-Canqui et al. (2011) found that for three long-term studies (>21 years) in Kansas, there was no significant increase in profile soil carbon to a 3.3 foot (1m) depth between no-till and conventional tillage management. Few additional studies have compared full-profile soil carbon content for Great Plains cropping systems. While the potential for increased carbon storage in soils is variable, the benefits of increased soil organic matter and improved soil structure associated with reduced tillage practices have been reported. In addition, the benefits of increased surface crop residue on reduced evaporation and temperature can allow cropping system diversification and intensification (Peterson and Westfall 2004, Westfall et al. 2010).

MULTI-FUNCTIONAL RANGELANDS AND PRAIRIES, LAND USE CHANGE

The large amount of native grazing lands and introduced pastures in the Great Plains landscape provide a multitude of ecosystem services in addition to agricultural production. They provide critical habitat for a number of species. However, fragmentation and degradation of the native vegetation through overgrazing, drought, and encroaching species, such as junipers, reduce the effectiveness of these lands for many species of concern, such as lesser prairie chicken, prairie dog, burrowing owls, and a wide variety of songbirds. Additionally, fragmentation and degradation of these grazing lands impede the hydrologic function of and nutrient cycling in the landscape.

Land use changes have slowed during the past 30 years (Parton et al. 2007), and recent analyses indicate the strong influence of conservation policies (Parton et al. 2007). The Conservation Reserve Program (CRP), from 1980 to 2000, allowed for less productive croplands to be taken out of production and converted back to grasslands thereby helping to reduce soil erosion and enhance biodiversity in agricultural landscapes. Recent conversion of CRP lands back to cropland is due to a combination of higher grain prices in response to the need for bioenergy feedstock and the termination of conservation contracts across the region. Other agricultural policy programs, including crop insurance, commodity, and disaster programs, also influence land use change from grasslands to croplands (Classen et al. 2011). In the future, additional feedstock production with second generation cellulosic bioenergy production technology may affect even larger areas of grassland environments. These climate and land use patterns present challenges and opportunities for grassland managers across the Great Plains.

Energy Resources

The Great Plains are rich with energy resources, from coal, oil, natural gas and nuclear, to wind, solar, biomass, biofuels, and geothermal. The extraction, transportation, processing, and sale of raw materials, fuels, and electricity provide jobs and incomes for communities throughout the region. However, there are challenges associated with these processes that will be exacerbated by the impacts of a changing climate. For example, large amounts of water are needed to produce natural gas and biofuels and to run power plants (see Chapter 6 for more on this topic). Higher average temperatures and drought will threaten water supplies and the operation of these facilities. Additionally, the increased flooding seen in recent years in the Great Plains also threatens power plants located in flood-prone areas.

Per capita energy consumption in the Great Plains is very high. It is the highest energy-consuming region in the United States, with Wyoming as the highest per capita energy consumption state consuming 956 million BTUs (1 million megajoules) per person compared to the national average of 308 million BTUs (0.3 million megajoules) (U.S. Department of Energy 2009). In addition, Texas and Wyoming are the biggest energy producers in the United States. Texas supplied 16.4%, primarily as natural gas and Wyoming contributed 14.23%, mostly as coal (U.S. Department of Energy 2009). At the same time, Texas has also been the fastest growing state for new wind energy facilities.

Choices about fuel portfolios will manifest differently for water and land resources. The nation has been moving away from coal-based electricity generation and toward natural gas over recent years. However, coal is unlikely to be removed from the fuel portfolio in the Great Plains as Wyoming's Powder River Basin is the largest producer of coal and provides the cleanest coal in the United States. Coal extraction in the West has increased in recent years as it has declined or remained stagnant in the eastern US. In addition, natural gas production in the West has increased and the Energy Information Administration projects continued growth. Both surface and sub-surface coal mining can have deleterious effects on the landscape and on water quality (Turka and Gray 2005). Assessment of the impacts of coal-bed gas development in the Powder River Basin found that impacts due to chemical spills and increased sedimentation into streams were potentially harmful to the health of fish and the riparian ecosystem as a whole (Farag et al. 2010).

The United States' production of oil and natural gas has increased dramatically, with shale oil and shale gas serving as the key driver. This has resulted in job growth in areas, like North Dakota, where the Bakken Shale discovery has unearthed oil reserves that are said to be more than Prudhoe Bay, Alaska. Increased production of natural gas has enabled natural gas prices to stay at record lows in the US. The tradeoff for these economic wins is increased carbon emissions and water quality impacts resulting from hydraulic fracturing ("fracking") in some areas. Major shale oil basins in the Great Plains include Bakken in North Dakota, Eagle Ford and Barnett in Texas, and Woodford in Oklahoma.

FUEL EXTRACTION AND WATER QUALITY

While data on the impacts of fuel extraction on water quality issues are only starting to emerge, this is an important area for future research as the risks of hydraulic fracturing to water quality and community health are increasingly becoming a significant public risk-perception issue, and conflict between local communities, the private energy sector, and government agencies is growing. One example of this is in the Williston Basin of North Dakota, Montana, and South Dakota. Information on the new research in this area can be found on the US Geological Survey site here: http://steppe.cr.usgs.gov/.

Exploration and extraction of fossil fuels for energy production can have major impacts on land use, ranging from vast surface mining, to the road networks connecting densely located well pads that blanket a landscape. Much of the new oil and gas production in the region relies on the method of hydraulic fracturing. This method of production employs diagonal drilling, which has limited some of the conversion on the land surface. However, well pads, storage infrastructure, and access roads can add to large changes in land use and land cover in certain regions. Figure 2.1 show the location of existing oil wells in the Williston Basin found in the northern portion of the Great Plains.

Additionally, hydraulic fracturing is a water-intensive production method, requiring anywhere from 2-9 million gallons (8-34 million liters) of water per site. The expansion of this industry has created new demands for water in a region that largely depends on groundwater from a diminishing aquifer. The financial income from oil and gas production are likely to exceed farm commodity revenues in many parts of the Great Plains and land use stands to be impacted as water rights are negotiated and change hands. States

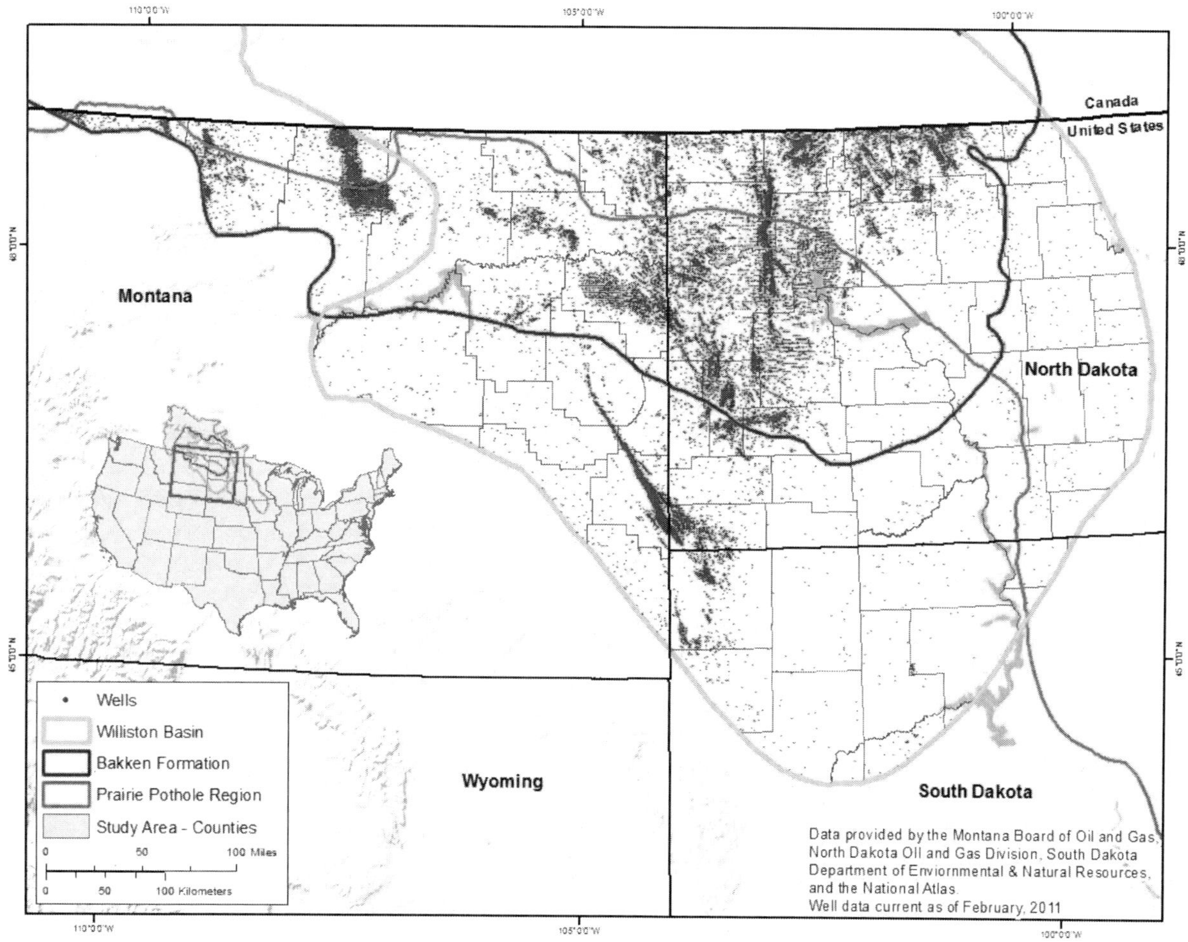

Figure 2.1. Map of the Williston Basin and Bakken Formation. Red points represent the spatial distribution of existing oil wells in the region. (Science Team about Energy and Prairie Pothole Environments 2011)

and municipalities are discussing ways to keep up with this industry's thrust. The city of Grand Prairie in the Barnett Shale of North Texas became the first municipality to ban the use of city water for fracking. Trucking water in from outside areas has started to take place, adding new energy demands to the production process.

There are multiple varying and uncertain factors that affect oil and gas well construction such as national and regional economic conditions, oil and gas prices, capital availability, corporate strategies, and technological innovations (COGCC 2012). In Colorado, the state agency, the Colorado Oil and Gas Conservation Commission, projects a 35% increase in water needed for hydraulic fracturing between 2010 and 2015 (COGCC 2012). The amount of water used depends on the geology of the region and whether wells are drilled horizontally or vertically, according to the Colorado Oil and

Gas Conservation Commission. Horizontal wells require more than vertical wells, as do shale oil and gas formations located deep underground versus shallower coal-bed methane sources.

WATER FOR POWER PLANT THERMOELECTRIC COOLING

The electricity generation system throughout the entire United States depends heavily on water for cooling. Wherever water scarcity is an issue, reliable production of electricity is also at risk. This is especially true in the western, drier portion of the country and the Great Plains region. Power plants built since 1980 typically use evaporative-cooling technologies that withdraw less, but consume more water. After the water is diverted from a local water body and used in the power plant, it is moved to a cooling tower or pond for reuse. This shift to evaporative-cooling technology is expected to continue, contributing to significant increases in energy-sector water consumption. In fact, the Electric Power Research Institute projected that 446 counties nationwide -- with the Southwest being hit especially hard -- would face water constraints on thermoelectric cooling by 2025, even if climate change has no effect on water supply. Looming water shortages are not the only threat that climate change poses for electricity generation. Many thermoelectric plants become less efficient on extremely hot days, when more energy needs to be expended on cooling the boiler water. Every part of the country is expected to see significant increases in hot days; many areas in the Great Plains are projected to have more than 75 days each year when the temperature tops 100°F (30°C), if climate change continues unabated (refer to Chapter 3 for climate information). Such hot days are typically when power plants have their peak demand as customers turn up their air conditioning. At the same time, the extreme heat can stress power system components, causing them to fail more quickly. Many transformers are designed to cool off at night and may be unable to cool down sufficiently. This design choice could be especially problematic because nighttime temperatures have been increasing faster than daytime temperatures (refer to Chapter 3 for climate information).

In North Dakota and Texas, thermoelectric power accounts for the most water withdrawals and represents 79% and 41%, respectively, of total withdrawals. In addition, looking at the magnitudes of withdrawals by state and sector, Texas' thermoelectric power withdrawals are the second largest in magnitude (10,800 thousand acre-feet per year (13322 thousand cubic-meter per year)) -- second only to Colorado's withdrawals for irrigated agriculture (11,200 thousand acre-feet per year (13815 thousand cubic meter per year)) (Kenny et al. 2009).

WATER AND LAND USE FOR RENEWABLE ENERGY SOURCES

The Great Plains are an ideal place for renewable energy production. The Great Plains states have a medium to high solar energy potential (National Renewable Energy Laboratory 2012a), and a fair to outstanding wind energy potential (National Renewable Energy Laboratory 2012b), and include many favorable sites for geothermal energy (National Renewable Energy Laboratory 2012c). In Colorado, the Department of Energy National Renewable Energy Laboratory reported that the state has considerable capacity for generating renewable energy through Photovoltaic installations on

Table 2.5 State in the Great Plains region with Renewable Portfolio Energy Standards (RPS) or Non-binding Goals

State	Minimum Amount of Renewables	Year	Organization Administering RPS
Colorado	20%	2020	Colorado Public Utilities Commission
Montana	15%	2015	Montana Public Service Commission
New Mexico	20%	2020	New Mexico Public Regulation Commission
North Dakota*	10%	2015	North Dakota Public Service Commission
South Dakota*	10%	2015	South Dakota Public Utility Commission
Texas	5.6 million BTUs per second (5,880 Megawatts)	2015	Public Utility Commission of Texas

*States with non-binding goals instead of RPS Source: (U.S. Environmental Protection Agency, 2012)
Useful resource: Database of state incentives for renewable and efficiency http://www.dsireusa.org/

non-irrigated farmland, which could contribute significantly to Colorado's Renewable Portfolio Standard (RPS) goals (Roberts 2011).

States in the Great Plains are putting RPS in place and using other mechanisms to build up the baseline of renewable energy sources (Table 2.5). An RPS specifies that electric utilities generate a certain amount of electricity from renewable or alternative energy sources by a given date. Nearly all of the Great Plains states have enacted an RPS (two have not), with goals ranging from 10% to 25%. Most of these are mandatory, with the exception of two states where the RPS is voluntary (Pew Center on Global Climate Change 2012). States in the Great Plains region that have RPS include: Montana, Texas, Colorado and New Mexico. North Dakota and South Dakota have nonbinding goals for renewable energy instead of an RPS. An RPS is a state requirement requiring electricity providers to obtain a minimum percentage of their power sources from renewable energy sources by a certain date.

However, water is a major constraint in the region to meet renewable energy production (Foti et al. 2011). The added impact of climate change will also negate any potential increase in Great Plains water availability due to increased water consumption across sectors (Foti et al. 2011). These results indicate the strong interaction between water usage among sectors in the Great Plains and the potential increase in productivity to agriculture and other socio-economic enterprises in the region (Foti et al. 2011). Ultimately, each type of energy use has influence on the environment, land use, and landscape conditions. Impacts result from extraction of requisite raw materials, transport

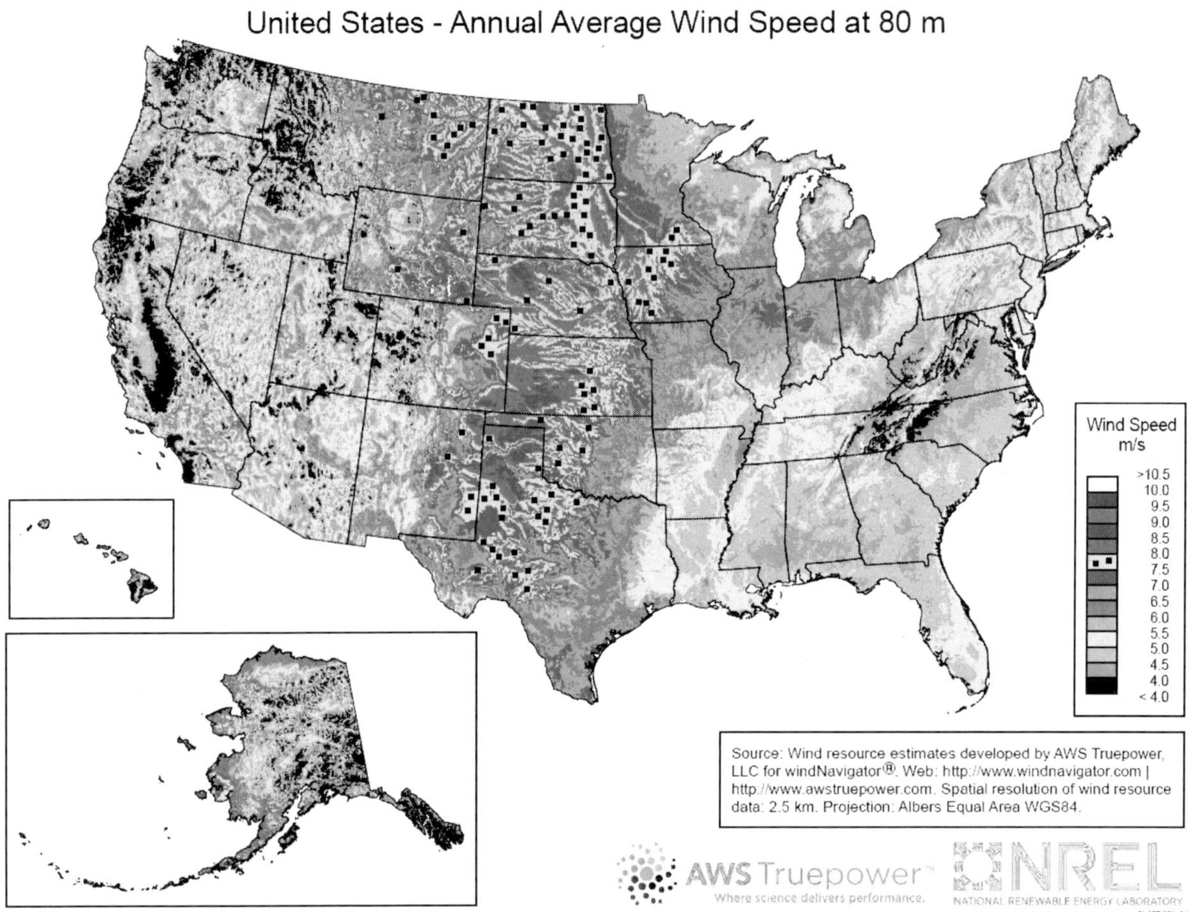

Figure 2.2. United States Annual Average Wind Speed at 80m, Source http://www.windpoweringamerica.gov/wind_maps.asp

from source to the production center to the end user, and any byproducts or end wastes produced. The availability and economic viability of energy choices can affect future land use and climate (Dale 1997). Demands for inputs, such as water, go hand and hand with energy and land-use decisions. Such demand requirements are likely to increase in the Great Plains as the region attempts to keep up with growing food, fiber, and energy developments.

Wind energy generation has expanded greatly across the Great Plains. While the resource inputs required for wind energy production are relatively small, infrastructure constraints associated to having access to transmissions lines have resulted in lower deployment of these wind systems on farming and grazing lands. Recent expansion of transmission lines has expanded the construction of wind energy facilities in the region. However, continued concern over building new roads and transmission lines to maintain the wind farms and transmit the generated energy, further fragmenting lands in non-cropped areas, which may additionally impact sensitive wildlife habitat.

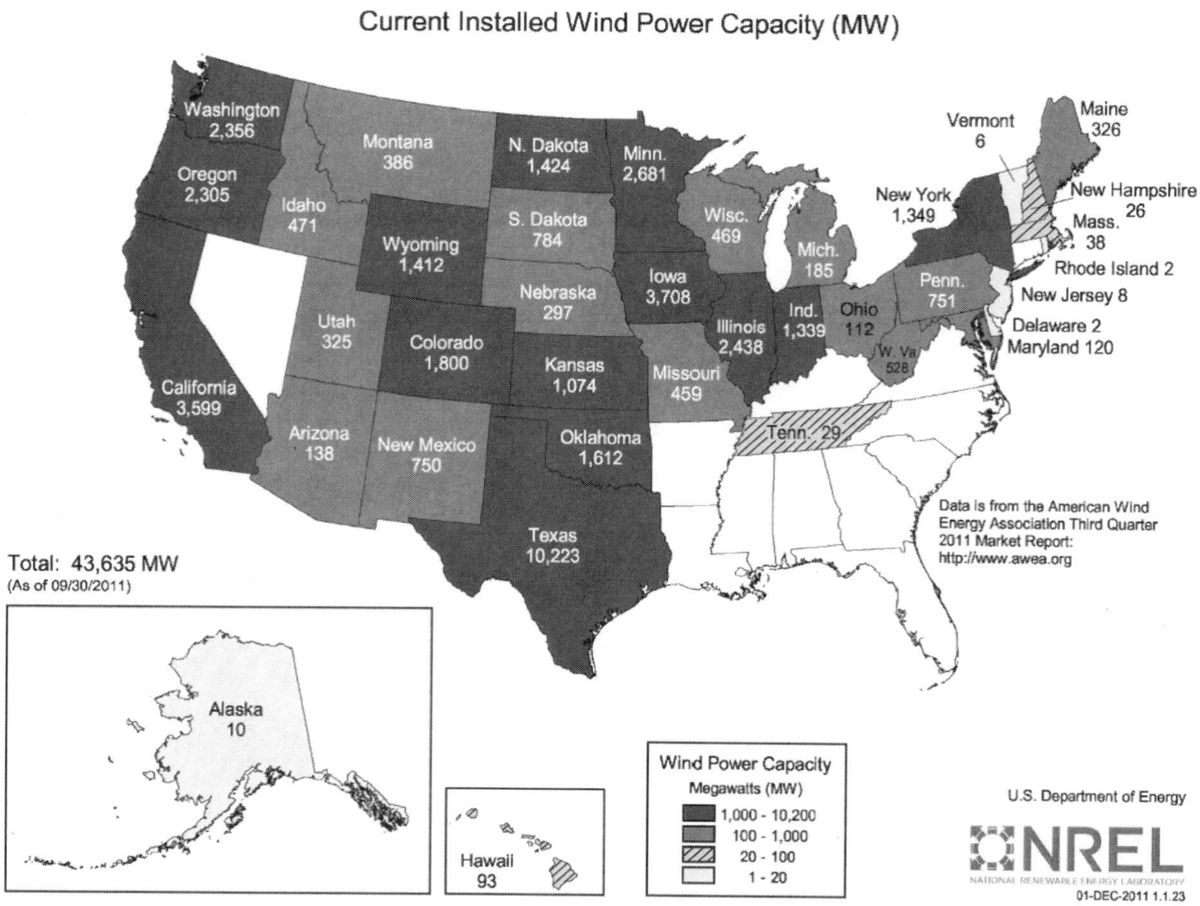

Figure 2.3. Current Installed Wind Power Capacity (MW), Source http://www.windpoweringamerica.gov/wind_installed_capacity.asp

The Great Plains region has the highest wind power capacity in the country. Texas is the state with the highest wind capacity built in 2011 (American Wind Energy Association 2011) and, as of this writing, Texas has by far the highest installed wind capacity of any state in the US, with 9.7 million BTUs per second (10,223 megawatts) (Figure 2.2).

The highest capacity (class 5) wind resource regions in the Great Plains can be found in the highlands of North Dakota and the high plains in Montana, while the next highest (class 4) exist in North and South Dakota, the Sandhills of Nebraska, northwest Oklahoma, south central Kansas, northeastern New Mexico, and the Texas Panhandle (Figure 2.3).

Resources for understanding the effects of wind energy development: http://www.fort.usgs.gov/WindEnergy/

© Danny Martin www.dannymartinphotography.com

Chapter 3

Climate Conditions and Scenarios of Change across the Great Plains

The Great Plains region experiences a wide range of extreme weather and climate events that affect society, ecosystems, and infrastructure. The large longitudinal range from North Dakota and Montana in the north to Texas in the south contributes to the extreme range in hot and cold temperatures. Climatic phenomena that have major impacts on the Great Plains include droughts, floods, winter storms, convective storms, heat waves, cold waves, hurricanes, and sea-level rise along the coastal area of Texas. The coastal regions are affected by storms reaching in the Gulf of Mexico and convective storms across the region can lead to heavy rainfall conditions throughout the Great Plains in the summer.

The Great Plains has a very wide range of annual average temperature (See Figure 3.1). The coldest temperatures of less than 40°F occur in the higher mountain areas of Wyoming and Montana and along the northern border with Canada. By contrast, the average annual temperatures in south Texas are greater than 70°F. Average annual precipitation (See Figure 3.2) also exhibits an extremely large range, illustrating the particular geographic features that determine the frequency of high moisture transport from oceanic sources. The far southeastern part of the region receives more than 60 inches per year, while some of the far western areas receive less than 10 inches per year.

Droughts across the Great Plains are frequent events and the region has experienced multi-year droughts. These droughts have been caused by high temperatures or by lack of rainfall, or both working in concert with each other. The 2011 drought in Texas and the southwest region of the Great Plains was one of most intense events during the past hundred years (NOAA 2011), and appears to be the most intense drought in the past 400 years.

Despite the low rainfall and high evaporative demand across most of the Great Plains, flooding events can and do occur in the region. These events reflect the temporal characteristics of episodic rainfall events associated with tropical depressions in the Gulf of Mexico, convective storms in the summers, and the rapid snowmelt occurring in the spring while soils may be saturated. The summer storms tend to be localized events associated with stationary convective storms moving slowly across the plains. The spring events have a larger regional impact, especially when spring snowmelt coincides with frontal weather patterns providing rainfall across a particular area. This type of event is similar to what occurred in 2011 floods along the Missouri River along with the extensive release from the dams in the upper reach of the Missouri system.

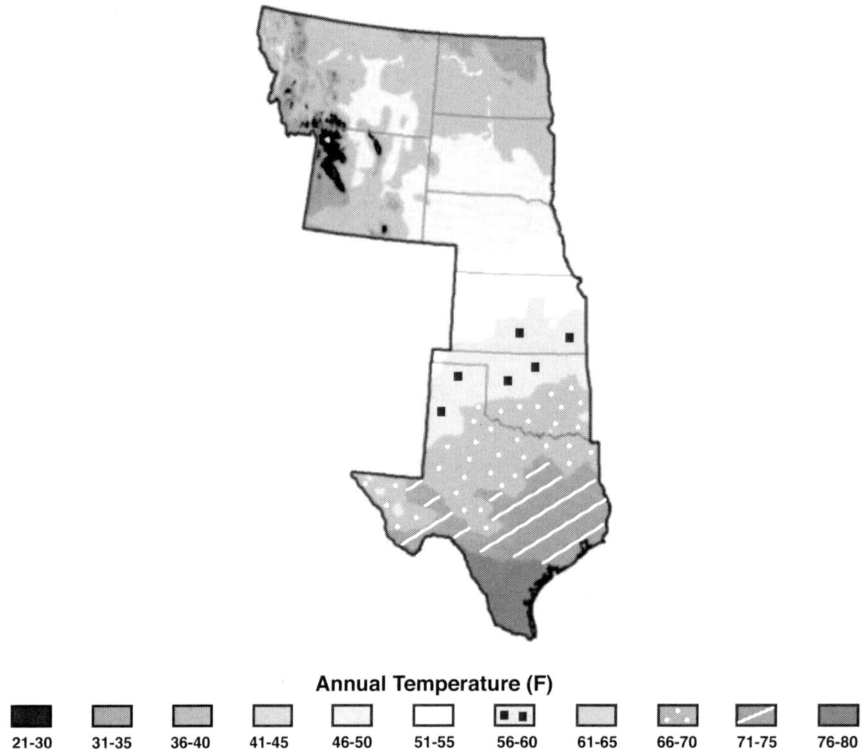

Figure 3.1. Average (1981 - 2010) annual temperature (°F) based on National Weather Service cooperative observer stations

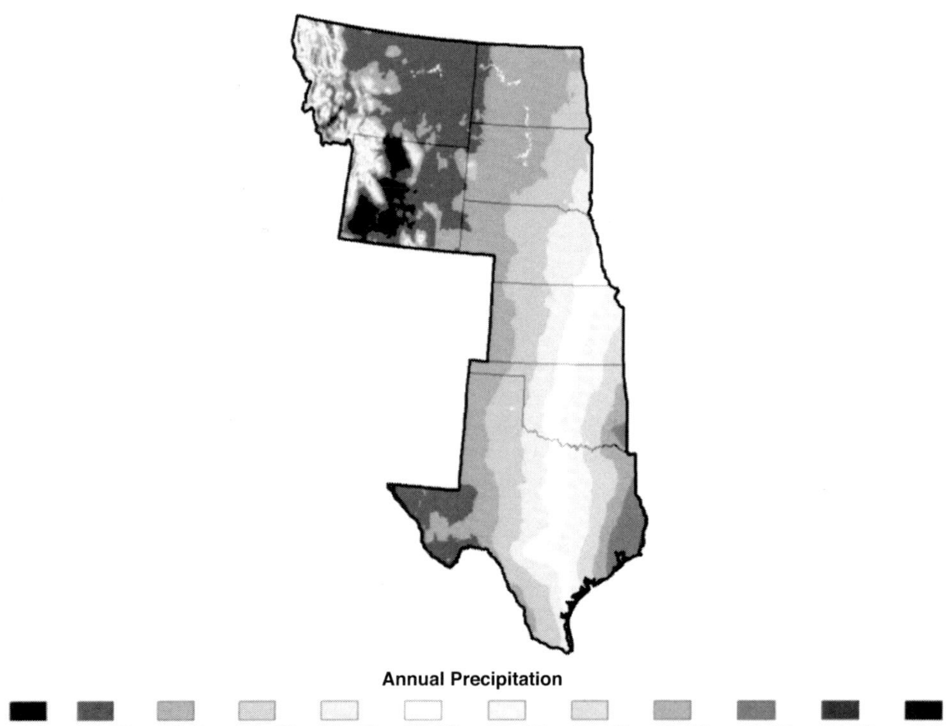

Figure 3.2. Average (1981- 2010) annual Precipitation (inches) based on National Weather Service cooperative observer stations

The region of the Red River of the North in North Dakota is a notable region of the northern Great Plains, in that it has been prone to flooding events due to a combination of increased precipitation and high levels of spring soil moisture. These conditions of coincident high soil moisture conditions and high spring rainfall in this low topographic valley during the spring thaw results in the flow of the river pooling up in shallow lakes with extensive flooding of these low lying areas.

The Great Plains is also prone to extreme winter storms, especially in the northern and central portions of the region. The polar jet stream can be found near or over the Great Plains during the winter months, bringing cold arctic air masses with the jet stream. The exposure to the winter jet stream dipping deeply into South Dakota and into Colorado and Kansas can lead to severe winter storms. These can also lead to ice storms as experienced in Oklahoma as snow transitions to rain in the southern Great Plains. These winter storms have an extensive impact on livestock, transportation, power lines, and human safety.

During summer months, differences in moisture levels and heating of the atmosphere can lead to extreme convective storms and to tornados. The atmospheric conditions of the Great Plains create these conditions and warm moist air moves in from the Gulf of Mexico and collides with relatively cool air moving along the jet stream. In May 2007, nearly 95% of Greensburg, KS, was completely destroyed by an EF5 tornado where 11 lives were lost. The event was part of a larger-scale tornado outbreak over a four-state region throughout the Plains.

In addition, to these intense convective systems and tornados the heat accumulation in the plains associated with high humidity levels can lead to heat stress events. These events can be lethal to people (Changnon et al. 1996, McGeehin and Mirabelli 2001) and livestock (Mader 2003, St. Pierre et al. 2003). Crops are also vulnerable to the heat stress conditions (Herrero and Johnson 1980).

Some examples of historic heat waves in the Great Plains region include the Dust Bowl of the 1930s (Schubert et al. 2004) and the 1980 summer heat wave and drought (Karl and Quayle 1981). Most recently, the heat wave and drought of the summer of 2011 across the southern portions of the Great Plains region had major impacts on human livelihood, crops, livestock, water supplies, and more.

Texas regularly experiences tropical storms and hurricanes. An extensive report on the climatology of hurricanes and tropical storms making landfall on the Texas coastline is found in Roth (2010). According to this report, the Texas coastline averages approximately 0.8 named storms per year. Roth (2010) also indicates that any given fifty mile coastal segment has an annual probability strike of approximately one storm per six years. Over the period of 1900 to 2010, these coastal areas have endured over 85 known tropical storms and hurricanes, the latter of which make up approximately half the events. As in other regions, the major impacts of tropical cyclones along the coast can be attributed to storm surge, high winds, and flooding from heavy rainfall.

The effects of hurricanes can extend well beyond the immediate coastal areas (Kruk et al. 2010). On occasion, the remnants of hurricanes will track northward and westward into the interior of the Great Plains. Such storms have caused heavy rainfall events from interior Texas to as far north as Nebraska. Over much of Oklahoma and interior Texas, between 3 and 6% of all days with more than 2 inches of rain result from these tropical cyclones (Knight and David 2009).

Climatic Trends

The temperature and precipitation data sets used to examine trends were obtained from NOAA's National Climatic Data Center (NCDC). The NCDC data is based on National Weather Service (NWS) Cooperative Observer Network (COOP) observations (Kunkel et al. 2013). Some analyses use daily observations for selected stations from the COOP network. Other analyses use a new national gridded monthly data set at a resolution of 5 x 5 km, for the time period of 1895-2011. This gridded data set is derived from bias-corrected monthly station data and is named the "Climate Division Database version 2 beta" (CDDv2) and is scheduled for public release in January 2013 (R. Vose, NCDC, personal communication, July 27, 2012).

Temperatures for the past 20 years have generally been above the 1901-1960 average, both annually and seasonally. Eight of the past ten summers (2002-2011) have been above the 1901-1960 average. The southern portion of the Great Plains experienced an extended period of hot days and drought summer of 2011 (See Figure 3.3). Northern states in the region have experienced the most change in their long-term average temperatures (e.g., North Dakota had the fastest increase in annual average temperature over the last 130 years, nationwide). Temperature trends are statistically significant (at the 95% level) for all seasons in the northern Great Plains and all seasons except summer and fall in the southern Great Plains.

Figure 3.3. Number of days exceeding 100 °F during the summer 2011.

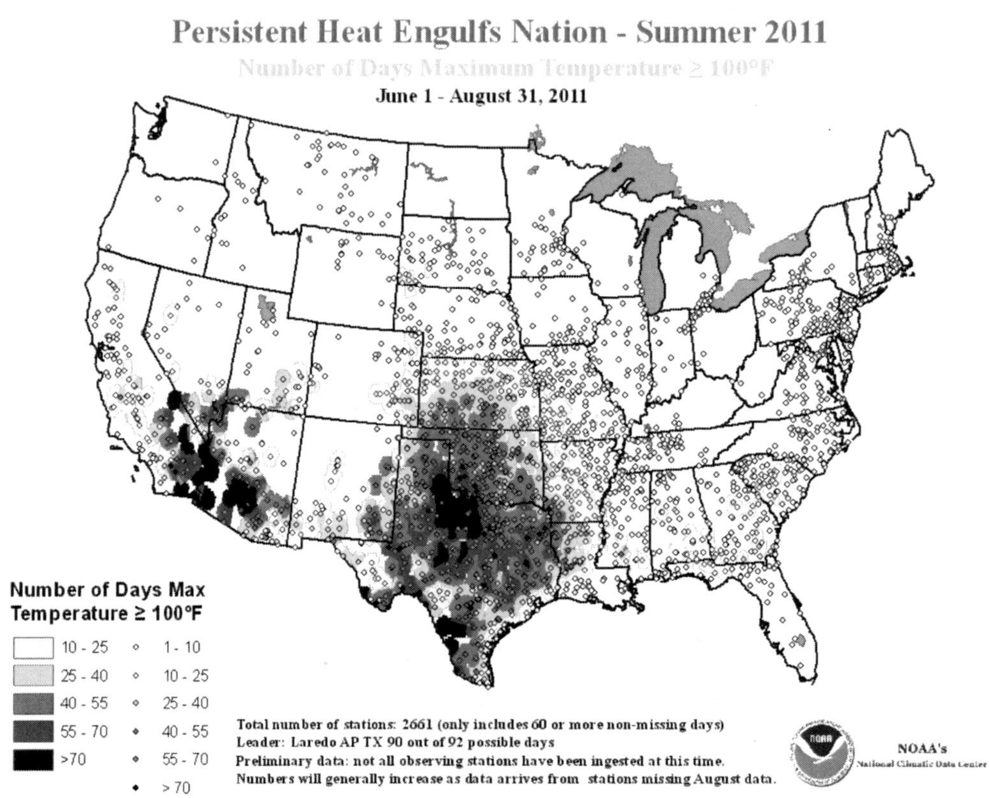

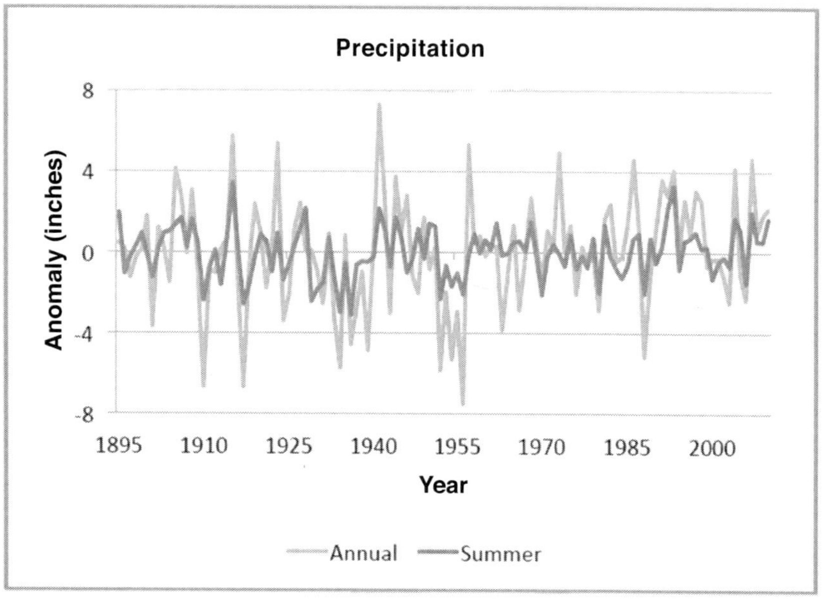

Figure 3.4. Precipitation anomaly (deviations from the 1901-1960 average, inches) for annual (black), winter (blue), spring (green), summer (red), and fall (orange), for the northern (solid lines) and southern (dashed lines) US Great Plains. Dashed lines indicate the best fit by minimizing the chi-square error statistic. Based on a new gridded version of COOP data from the National Climatic Data Center, the CDDv2 data set (R. Vose, personal communication, July 27, 2012). Note that the annual time series is on a unique scale. Trends are not statistically significant for any season.

Trends in precipitation are not statistically significant. Although, for the 1990's annual precipitation for the Great Plains was greater than normal, during the early 2000's less than normal, and greater than normal during the last few years except for 2011. Occurrence of extreme (heavy) precipitation events also exhibits substantial inter-annual and decadal-scale variability. Since 1990, there have been several years with a very high number of extreme precipitation events often associated with tropical cyclones, with the greatest overall value of 1-day events occurring in 2007 (Kunkel et al. 2010, 2013).

Extreme cold and hot periods exhibit a large amount of inter-annual variability. The recent tendency toward fewer extreme cold events is more prominent in the north than in the south. Historical occurrence of extreme hot events is dominated by the severe heat of the 1930s. There has been a generally increasing trend in freeze-free season length since the early 20th century. The average freeze-free season length during 1991-2010 was about 6 days longer than during 1961-1990.

Figure 3.4 shows annual and seasonal time series of precipitation anomalies for the period of 1895-2011, for both the northern and southern Great Plains calculated using the CDDv2 data set. The variability of precipitation is greater in the southern Great Plains than in the north. Annual precipitation for the entire Great Plains region was greater than the 1901-1960 average during the 1990s, less than the average during the early 2000s, and greater than the average during the last few years, except for 2011. The early 1950s were the driest multi-year period, and included the single driest year on record, 1956. The 1930s were nearly as dry. The wettest single year on record was 1941. Summer precipitation anomalies are very similar to the annual behavior, except that the 1930s were the driest multi-year period. In fact, the driest summer on record is 1936 for Oklahoma, Kansas, Nebraska, South Dakota, and North Dakota. The flood year of 1993 was the second wettest summer on record, after 1915. The severe impacts of the 1930s in the Great Plains can be attributed mainly to the conditions during the summers, which were much more severe than during the multiyear dry period of the 1950s. For the region as a whole, 1934 and 1936 were the two hottest summers on record and the two driest summers on record. This combination of heat and dryness, along with the close temporal proximity of these two extreme summers, is unique in the record.

There is no overall trend in the occurrence of heat waves. The frequency of extreme cold periods has been generally low since 1990 (averaging about 65% below the long-term mean), with the exception of 1996 when a severe cold wave in early February affected large areas. Other recent years with widespread severe cold included 1983 and 1989. The 1950s were a period of few severe cold waves (averaging about 60% below the long-term mean). A separate analysis of the northern and southern parts of the region indicates that the recent tendency toward fewer cold waves is more prominent in the north than in the south.

Simulated Climate Scenarios

This section summarizes climate model simulations for two scenarios of the future path of greenhouse gas emissions: the IPCC SRES high (A2) and low (B1) emissions scenarios. These simulations incorporate analyses from multiple sources, the core source being Coupled Model Inter-comparison Project 3 (CMIP3) simulations. Additional sources consist of statistically- and dynamically-downscaled data sets, including simulations from the North American Regional Climate Change Assessment Program (NARCCAP). Analyses of the simulated future climate are provided for the periods of 2021-2050, 2041-2070, and 2070-2099), with changes calculated with respect to a historical climate reference period (1971-1999, 1971-2000, or 1980-2000). The resulting climate conditions are to be viewed as scenarios, not forecasts, and there are no explicit or implicit assumptions about the probability of occurrence of either scenario. The basis for these climate scenarios (emissions scenarios and sources of climate information) were considered and approved by the National Climate Assessment Development and Advisory Committee.

Climate analysis of the effect of increased emissions of warming gases into the atmosphere simulated similar spatial patterns of annual temperature increase. The higher emission (A2) scenarios simulated greater warming for the northeastern portion of the Great Plains. These models all indicated a significant warming across the Great Plains for both emission scenarios. The CMIP3 scenarios for the high emission (A2) simulations average temperature increases of 2.8°F by 2035, 4.4°F by 2055, and nearly 8°F by

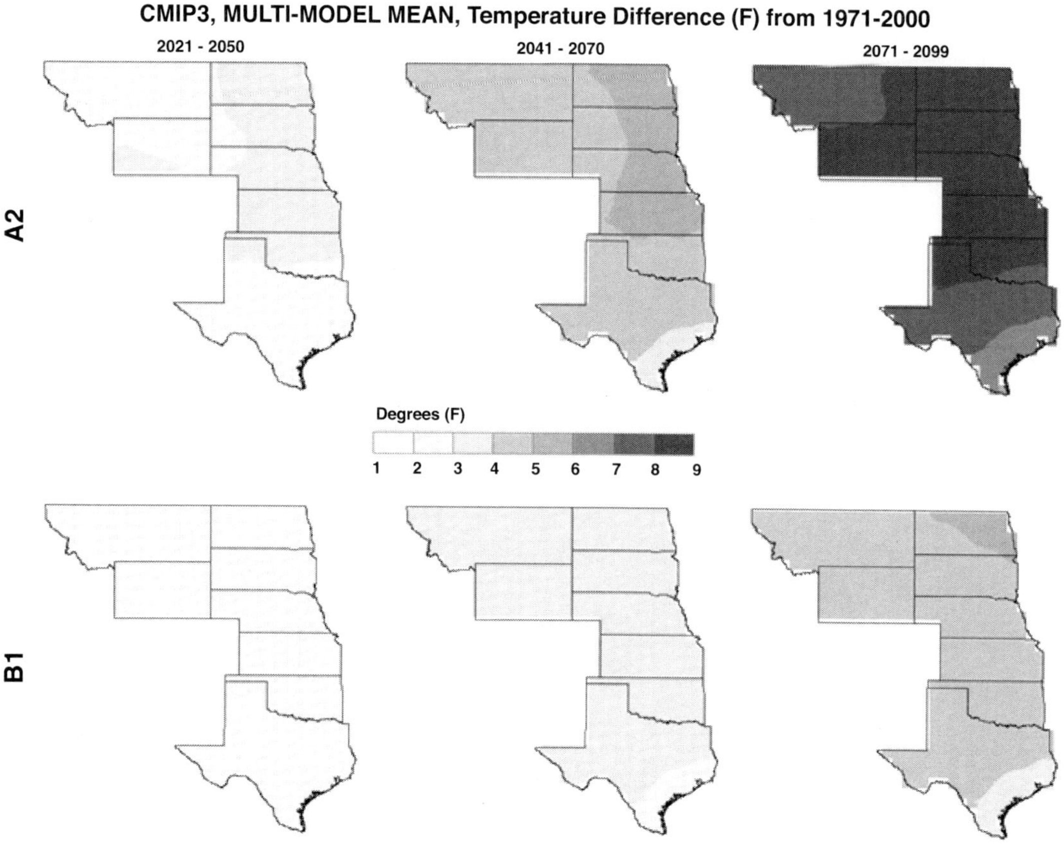

Figure 3.5. Mean annual temperature changes (°F) for each future time period with respect to the reference period of 1971-2000 for all 15 CMIP3 models, averaged over the entire Great Plains region for the high (A2) and low (B1) emissions scenarios. Also shown are results for the NARCCAP simulations for 2041-2070 and the 4 GCMs used in the NARCCAP experiment (A2 only). The small plus signs are values for each individual model and the circles depict the overall means.

2085 for the Great Plains (See Figure 3.5). These increases are statistically different from the decade of 2001 to 2010 for the high emission scenario by mid-century and beyond. Simulated increases under the low (B1) emissions scenario are very similar to those of the A2 scenario by 2035, but are considerably smaller by 2085. These low (B1) emission scenario simulations of average annual temperature differences do not show significant differences until the latter part of the century (2085). Seasonal temperature increases are simulated to be largest in summer and smallest in spring.

Regional differences in simulated increase in the number of days exceeding 95°F were calculated from the NARCCAP results (Figure 3.6). The analysis indicated the largest increases (more than 30 days) to occur in the southwest corner of Texas. The simulations also indicated increases in the number of consecutive days above 95°F and this region of Oklahoma and Texas was calculated to have 12 days or more days above 95°F. In the

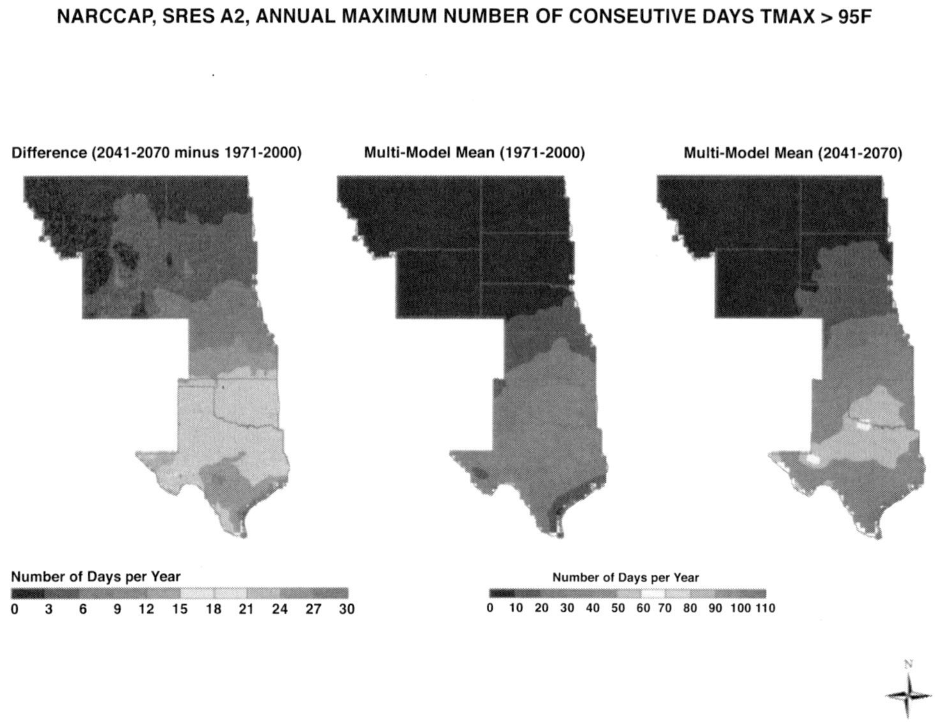

Figure 3.6. Spatial distribution of the NARCCAP multi-model mean change in the annual maximum number of consecutive days with a maximum temperature greater than 95°F between 2041-2070 and 1971-2000 (top). Model reference periods of the annual maximum number of consecutive days with a maximum temperature greater than 95°F (bottom).

central and northern portions of the region, the changes are smaller, generally in the range of increases of 4-12 days. Regionally, the smallest increases (less than 10 days) were seen in the far northern portion of the region in high elevation areas of Montana. Across the far northern tier of the region, the increase in the number of consecutive days exceeding 95°F is less than 4 days (for the A2 scenario at mid-century).

The simulated seasonal warming effects on winter temperatures indicated that the greatest reduction in the number of days with a minimum temperature below 32°F (more than 28 days) occurred in the northwestern part of the region (for the A2 scenario at mid-century) (Figure 3.7). Overall, the freeze free season simulated becomes longer throughout the region. Increases in freeze-free season are approximately 20 to 30 days longer across the region by mid-century for the high emission (A2) scenario. Simulated growing degree days (base temperature set at 50°F) over the region was calculated to increase by over 25% by mid-century (Figure 3.8).

Modeling of precipitation still proves to be difficult with model simulations displaying wide ranges of variation around the average of the simulation results. However, the models indicate a general decrease in average annual precipitation in the southern areas of the Great Plains and increase in precipitation in the northern areas.

Regional patterns of precipitation intensity indicate that over the entire Great Plains, simulated precipitation days which exceed 1 inch increases up to 27% for the high emission (A2) scenario by mid-century. The western portion of the region is more uncertain of how these number of 1 inch precipitation days will be affected. As for days

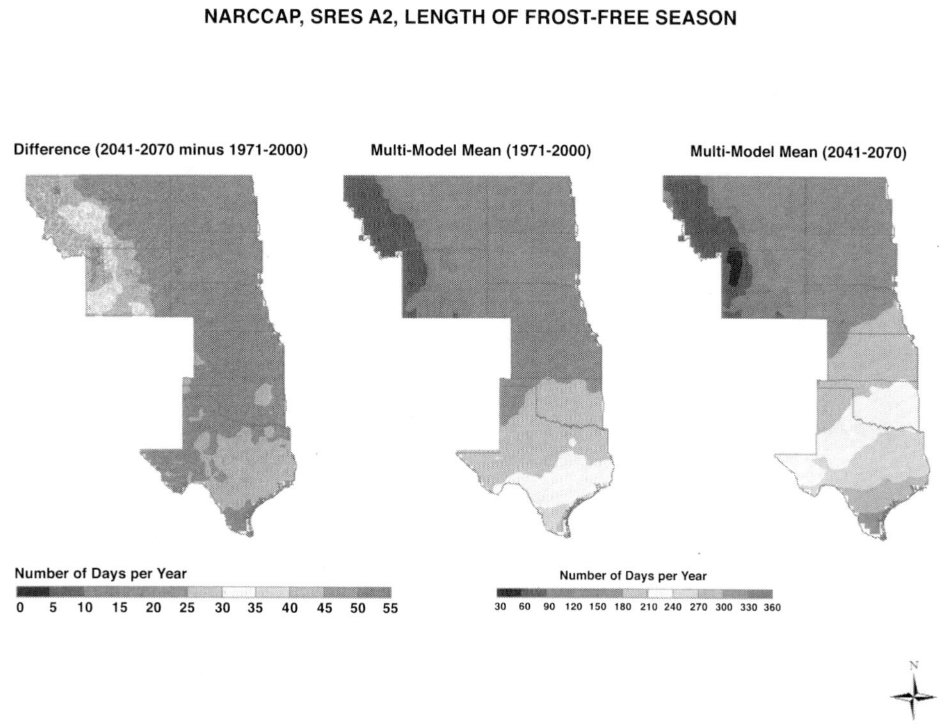

Figure 3.7. Spatial distribution of the NARCCAP multi-model mean change in the length of the frost-free season between 2041-2070 and 1971-2000 (top). Model reference periods of the length of the frost-free season (bottom).

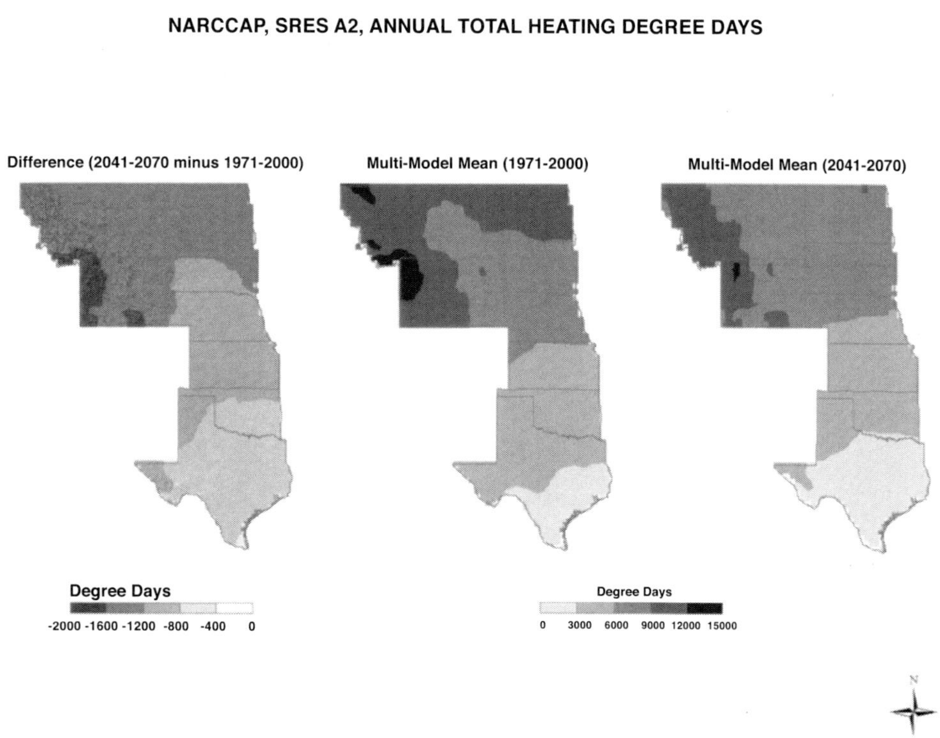

Figure 3.8. Spatial distribution of the NARCCAP multi-model mean change in the annual maximum number of consecutive days with a maximum temperature greater than 95°F between 2041-2070 and 1971-2000 (top). Model reference periods of the annual maximum number of consecutive days with a maximum temperature greater than 95°F (bottom).

with little precipitation (less than 0.1 inches), simulated results from the high emission (A2) scenario indicates that in the south, the number of days of low precipitation by 3 to 13 days per year by mid-century would increase. In the north, the opposite response is simulated under the high emission scenario with up to 8 days per year less occurrence of low precipitation days.

SECTION 2

Natural Resource Vulnerabilities and Challenges Faced by the Great Plains

The changing environmental factors faced by the Great Plains and its residents will affect social and economic activities in the near and long term. Recent trends have concentrated populations in more urban centers with rural areas still providing significant economic development through agricultural production.

Climate change will affect water availability and other environmental elements as well as energy production in the Great Plains, and also test community infrastructure and current land management impacting both economic and ecological health. A change or increase in the frequency and intensity of extreme weather events pose a particular risk to human and environmental systems, including agriculture, water resources, energy development, biodiversity and wildlife. Residents, land managers and government officials can plan for changes through mitigation and adaptation measures, which may require major shifts in individual and institutional practices and mindsets.

© Amber Childress

Chapter 4

Water Management

Water in the Great Plains is a critical natural resource that determines the social-ecological processes related to conservation, agriculture, energy, and urban development, among others. Climate regimes across the Great Plains vary tremendously and affect seasonal distribution of water inputs and availability. Changes in precipitation patterns, such as the variability and intensity of rain or snowfall, and seasonality of precipitation have major impacts on water resources in the region. In addition, the river systems dissecting the Great Plains, such as the Red River of the North, the Missouri, the Platte, the Arkansas, and the Rio Grande basins, emerge from the Rocky Mountains, so the hydrologic flow is connected to the snow deposition in this region. This is complicated by a legal allocation system that determines when, where, and how much water can be diverted and used in the region. Determinants of these allocation rules were developed during the past century and evolved under more ample precipitation conditions; and when population was sparser; industrial, energy, and urban demands were lower; and environmental water flow requirements were of lower priority. Water usage across the Great Plains is dominated by agriculture demands, though increased concentrations of regional urban development have affected water rights and usage. Changes in water ownership during the past few decades have also caused increased transfer of water rights to various municipalities. This has resulted in conflicts and legal battles between states and between various uses and users.

Local water development has been augmented greatly over the decades through development of diversions and reservoirs (primarily public investment) and the drilling of wells into aquifers (large private investment as well as public). These water infrastructure developments have altered stream and river flows, wetland extent, hydrological dynamics, and sedimentation rates that affect river and stream morphology and reservoir storage capacity. Climate scientists predict that water cycles will be altered so that the past precipitation patterns no longer provide a guide for the future (Milly et al. 2008). This will require new ways to manage and govern water resources in the context of all the multiple climatic and non-climatic stressors involved (Ison et al. 2007, Steyaert and Jiggins 2007, Pahl-Wostl 2007, Norgaard et al. 2009, Birkmann et al. 2010, Lebel et al. 2010, Farrelly and Brown 2011, van de Meene et al. 2011, Huntjens et al. 2012).

Water Use and Management

Multiple and diverse users compete for water in the Great Plains region. Agriculture, however, is by far the biggest user of water, accounting for 65% of combined fresh water withdrawals (Kenny et al. 2009). Other uses include urban and rural domestic and municipal entities, energy extraction and power production, industry, recreation, and wetlands and riparian ecosystems, as well as aesthetic and spiritual uses. Thermoelectric power and public supply account for 21% and 10% of Great Plains water withdrawals, respectively. In North Dakota and Texas, thermoelectric power accounts for the majority of withdrawals, 79% and 41%, respectively (Kenny et al. 2009). In Oklahoma, public water supply (42%) is the largest user (Kenny et al. 2009). Maintaining ecosystems services provided by water and the well-being of all life that depends on clean and available water requires careful management and policies to sustain adequate water quality and quantity in a variable and changing climate (Rosenzweig et al. 2004).

When considering fresh surface and groundwater sources separately, surface water supplies 68% of Great Plains water needs and groundwater provides 32% (Table 4.1). For irrigated agriculture, surface water provides 57% and groundwater 43% of total withdrawals. However, at a state level, the distribution is more skewed. In Colorado, Montana, and Wyoming, surface water provides over 80% of irrigation needs. In Kansas, Nebraska, and Texas, groundwater provides over 75% of irrigation needs.

Groundwater Issues as They Relate to Climate Variability

Water level changes in the High Plains (Ogallala) Aquifer from the time prior to extensive groundwater irrigation (before about 1950) to 2009 ranged between a rise of 41 feet (12.5 m) and a decline of 178 feet (54.3 m) with an average water-level decline of 14 feet (4.3 m) since predevelopment (McGuire 2011). Total storage of the Ogallala Aquifer has declined by 274 million acre-feet (333,040 million m^3) since predevelopment (McGuire 2011). Groundwater withdrawals from the High Plains Aquifer in 2000 accounted for 20% of the total US groundwater withdrawn, 97% of which is used for irrigation (Maupin and Barber 2005). Groundwater extraction for drinking water supports about 82% of the people in the High Plains aquifer region (Gurdak et al. 2011). Groundwater from the vast Ogallala Aquifer in the Central Plains, one of the largest aquifers in the world, is predicted to continually decline as long as irrigation remains viable given escalating pumping costs and overall farm production costs for seed, fertilizer, equipment, and other related expenses (Howell 2009). Water right transfers from agriculture to urban and industrial uses will further exacerbate this inevitable resource strain. Weather directly affects the water requirements of crops and, thus, their irrigation requirements (Howell 2009). An indirect effect of climate change is increased groundwater pumping, which could affect hydraulic heads in many aquifers, allowing upward leakage of groundwater with poorer water quality, such as in the High Plains aquifer (McMahon et al. 2007).

Groundwater depth determines regions' relative susceptibility to precipitation and temperature changes, and groundwater storage acts as a moderator of watershed response and climate feedbacks (Maxwell and Kollet 2008). There is a "critical zone" of groundwater depth – between 7 to 16 ft (2 to 5 meters) deep – where there is a very strong correlation between water-table depth and surface evaporative demand (Maxwell and Kollet 2008). Playa lakes are unique hydrological formations to the High Plains area and essential for recharging the Ogallala Aquifer, which means they play an important role in groundwater management and aquifer sustainability (Gurdak and Roe 2010). There are approximately 61,000 playas in the region, with the highest concentration in the

Table 4.1 Total surface water and groundwater withdrawals in the Great Plains region by state and water-use category in 2005, in thousand acre-feet per year (3.1a) and in thousand cubic meters (3.1b) (values may not sum to totals because of independent rounding)

	Public		Domestic		Irrigation		Livestock		Aquaculture		Industrial		Mining		Thermoelec		Total	
State	SW	GW	SW	GW	SW	GW	SW	GW	SW	GW	SW	GW	SW	GW	SW	GW	SW	GW
Colorado	855	114	0	39	11,200	2,600	12	25	80	91	156	4	1	6	131	7	12,400	2,810
Kansas	272	180	0	17	128	2,940	27	95	4	2	7	40	5	11	499	15	942	3,300
Montana	84	75	1	25	10,700	157	31	13	44	3	33	42	38	1	100	0	11,00	317
Nebraska	106	264	0	58	1,290	8,190	23	99	83	10	0	13	11	0	3,970	9	5,480	8,650
North Dakota	39	36	0	10	82	87	10	15	7	0	11	6	0	6	1,190	0	1,340	160
Oklahoma	597	127	0	28	150	405	120	61	21	0	18	9	2	1	183	1	1,090	634
South Dakota	39	74	0	9	160	167	32	22	16	21	0	5	7	5	4	1	258	303
Texas	3,440	1,350	0	288	1,890	6,860	108	182	10	6	1,190	210	72	30	10,800	63	17,500	8,990
Wyoming	52	56	0	7	4,000	474	11	7	24	3	2	5	15	43	248	1	4,350	595
GP Totals	5,484	2,276	1	481	29,600	21,880	374	518	289	64	1,417	333	152	104	17,125	98	54,360	25,759
Colorado	855	114	0	39	11,200	2,600	12	25	80	91	156	4	1	6	131	7	12,400	2,810
Kansas	272	180	0	17	128	2,940	27	95	4	2	7	40	5	11	499	15	942	3,300
Montana	84	75	1	25	10,700	157	31	13	44	3	33	42	38	1	100	0	11,00	317
Nebraska	106	264	0	58	1,290	8,190	23	99	83	10	0	13	11	0	3,970	9	5,480	8,650
North Dakota	39	36	0	10	82	87	10	15	7	0	11	6	0	6	1,190	0	1,340	160
Oklahoma	597	127	0	28	150	405	120	61	21	0	18	9	2	1	183	1	1,090	634
South Dakota	39	74	0	9	160	167	32	22	16	21	0	5	7	5	4	1	258	303
Texas	3,440	1,350	0	288	1,890	6,860	108	182	10	6	1,190	210	72	30	10,800	63	17,500	8,990
Wyoming	52	56	0	7	4,000	474	11	7	24	3	2	5	15	43	248	1	4,350	595
GP Totals	5,484	2,276	1	481	29,600	21,880	374	518	289	64	1,417	333	152	104	17,125	98	54,360	25,759

4.1a. 2005 fresh water withdrawals by water-use category in thousand acre-feet per year separated by surface and groundwater sources

Source: (Kenny et al., 2009)

Table 4.1 Total surface water and groundwater withdrawals in the Great Plains region by state and water-use category in 2005, in thousand acre-feet per year (3.1a) and in thousand cubic meters (3.1b) (values may not sum to totals because of independent rounding)

	Public		Domestic		Irrigation		Livestock		Aquaculture		Industrial		Mining		Thermoelec		Total	
State	SW	GW	SW	GW	SW	GW	SW	GW	SW	GW	SW	GW	SW	GW	SW	GW	SW	GW
Colorado	1,055	141	0	48	13,815	3,207	15	31	98	23	192	5	2	7	162	9	15,295	3,466
Kansas	336	22	0	21	158	3,626	33	117	4	3	9	49	6	14	616	19	1,162	4,070
Montana	103	93	1	31	13,198	194	38	16	55	3	41	52	47	2	123	0	13,568	391
Nebraska	131	326	0	72	1,591	10,102	28	122	103	12	0	16	14	0	4,897	11	6,759	10,670
North Dakota	49	44	0	12	101	108	12	19	9	0	13	7	1	7	1,468	0	1,653	197
Oklahoma	736	157	0	35	185	500	148	76	26	0	22	11	2	1	226	2	1,344	782
South Dakota	48	91	0	11	197	206	39	27	19	26	0	6	8	6	5	1	318	374
Texas	4,243	1,665	0	355	2,331	8,462	133	224	13	7	1,468	259	89	37	13,322	77	21,586	11,089
Wyoming	64	69	0	9	4,934	585	14	8	29	3	3	6	19	53	306	2	5,366	734
GP Totals	6,765	2,807	1	593	3,6511	26,989	461	639	356	78	1,748	410	188	128	21,124	120	67,052	31,773

4.1b. 2005 fresh water withdrawals by water-use category in thousand cubic meters per year separated by surface and groundwater sources

southern region in Texas and part of the central and northern High Plains aquifer region in Kansas and Nebraska (Gurdak and Roe 2009, 2010). New techniques monitor surface and sub-surface groundwater levels using the Gravity Recovery and Climate Experiment satellite, which uses gravity to measure groundwater, soil moisture, surface water, snow and ice, and biomass. These new practices will become increasingly important for understanding how to manage for irrigation and sustainable agroecosystems and the relative influences of climate change versus agricultural practices (Strassberg et al. 2009, Scanlon et al. 2010).

Water quality of the High Plains Aquifer will be impacted by decreased precipitation or drought due to increased groundwater pumping from high-capacity wells, thereby increasing upward flow of saline groundwater from underlying geologic units and further reducing groundwater quality (Gurdak et al. 2011). The combined effects of groundwater development and climate change may also lead to less dilution of contaminants in streams during low flows than was assumed in setting stream-discharge permits (Alley 2001, Green et al. 2011). Climate variations associated with the Pacific Decadal, Atlantic Multi-decadal Oscillation, and El Niño Southern Oscillation, have been identified as having substantial control on the recharge and water-table fluctuations on the High Plains Aquifer (Gurdak et al. 2007, McMahon et al. 2007, Green et al. 2011).

Across regions of the High Plains Aquifer in Kansas, streamflow declines were historically caused by high rates of groundwater pumping, but also correlate with climate variability since the mid- 1980s (Brikowski 2008). Brikowski (2008) showed that projected climate change for the region will likely continue streamflow declines, resulting in severe consequences for surface-water supply and the strong possibility of unsustainable surface storage of water resources in the region. The result may lead to greater pressure on the groundwater resources of the already- stressed High Plains Aquifer (Brikowski 2008).

In southeastern Colorado, salinization and degradation of both groundwater and surface water through excessive irrigation and seepage have occurred, which can lead to diminishing crop yields (Gates et al. 2006). All of these changes to groundwater and surface water quantity and quality are increasingly critical because they remove risk buffers of climate change impacts on water availability.

Evapotranspiration influences the amount of water ultimately reaching rivers, and it affects the amount of water needed for irrigation in agricultural areas. Changes in temperature magnitudes, precipitation amounts and timing, and CO_2 concentrations will all influence evapotranspiration, sometimes in conflicting ways. Precipitation and temperature changes could act in combination either to enhance plant growth, which could increase total amounts of evapotranspiration occurring, or they could act to decrease plant growth; for instance, if a species' optimal temperature range was exceeded, which could decrease plant growth and evapotranspiration (Thomson et al. 2005, Spears et al. 2011).

Increased CO_2 concentrations also affect plants and evapotranspiration. Some studies have shown that higher CO_2 concentrations may lead to increases in leaf area and plant growth and vigor, which could lead to increased ET and water consumption overall (Baldocchi and Wong 2006, Spears et al. 2011, Wu et al. 2012). On the other hand, under higher CO_2 conditions, plants have been observed to partially close the stomatal openings on their leaves, which results in decreased transpiration and water loss

(Field et al. 1995, Gedney et al. 2006), possibly because higher concentrations mean that less stomatal opening is required for plants to absorb the amount of CO_2 they need for photosynthesis (Sellers 1996, Wu et al. 2012). One study provides evidence suggesting that the rise in continental runoff observed during the 20th century is consistent with CO_2-induced suppression of transpiration (Gedney et al. 2006). In snowmelt-dominated regions, such as the Rocky Mountains, which are the headwaters for many of the Great Plains' rivers, snowmelt earlier in the season would result in increased soil moisture at a time when potential evaporation is lower than has historically been the case (Barnett et al. 2005).

In contrast to evapotranspiration, which includes vegetative water losses, evaporation may also take place directly from water surfaces of streams and reservoirs. Reservoir evaporation in the Great Plains is currently considerable. For instance, annual evaporation from the six largest reservoirs on the Missouri River's main stem has been estimated to be about 5% of the average annual river discharge (Benke and Cushing 2005). In the Rio Grande, evaporation from the major reservoirs has been estimated to exceed municipal water usage in the basin . Such reservoir losses could increase if warmer temperatures dominate other factors.

Water Resources and Climate Change Projections

Water demand across the Great Plains associated with the A1B and B2 (see Chapter 3 regarding climate scenarios and Kunkel et al. 2013 used for this report) climate projections indicate that the central portion will experience a slight decline in water yields, ranging up to 1.2 in/yr (3 cm/yr) decline (Foti et al. 2011). Western Montana and Wyoming will potentially be affected by lower water yields across a set of climate projections, ranging from 0.8 to 3.1 in/yr (2 to 8 cm/yr) decreases (Foti et al. 2011). The projections for the eastern fringe of the Great Plains indicate a consistent decline in water yields (Foti et al. 2011). Southern Texas demonstrates the greatest variability in water yields for this portion of the Great Plains, associated with model characteristics providing the specific rainfall pattern (Foti et al. 2011).

Evidence suggests that the Missouri River Basin as a whole may have experienced relatively wetter conditions during the 20th century compared to prior centuries as well as relatively less annual runoff variability (U.S. Bureau of Reclamation 2011). Even omitting major flood events in 1996 and 1997, the 1990s were still the sixth wettest decade of the past 300 years (using data from the Yellowstone River) (Graumlich et al. 2003, U.S. Bureau of Reclamation 2011). Climate reconstructions, based on tree-ring data, have indicated that the 1930s were the driest extended period during the past 300 years with below average stream flows and the 1930's drought was virtually unprecedented during this 300-year record (U.S. Bureau of Reclamation 2011). Observations from 1957 to 2007 across 202 gauging stations in the Missouri River Basin indicate that stream flows are down in the western part of the basin and up in the eastern part (Anderson et al. 2008).

Elgaali and Garcia (2007) found that there are already shortages in surface water supply in the Colorado portion of the Arkanas River Basin, and a small amount (5% to 10%)

of these shortages is met by groundwater pumping. The analysis used two different climate scenarios based on the VEMAP climate data sets (Kittel et al. 1995), which generated a statistical downscaled product, approximately 0.5 degrees spatial resolution and monthly and daily data products for temperatures and precipitation, using climate projections from the Canadian Climate Center (CCC) and the Hadley Center (HAD) model results. Analysis based on the CCC projection suggests that the region could experience a shortage in water supply from the 2010s to the 2090s for the whole season and for each month from May through September, with the summer facing greater shortages than the spring. Results based on the HAD projection also suggest a shortage in August, but with no shortages over the whole season, assuming that there is sufficient storage in the system to hold water.

Climate projections for the Missouri River Basin as a whole (i.e. at Omaha), indicate that the mean annual temperatures for the 2020s, 2050s, and 2070s decades will be 1.6 °F (0.9 °C), 3.5 °F (1.9 °C), and 4.8 °F (2.7 °C) higher, respectively, than that for the 1990s (U.S. Bureau of Reclamation 2011). The ensemble median shows a gradual increase in basin-wide precipitation over the 21st century – up to an 8.5% increase by the 2070s, as compared to the 1990s. However, individual projections are not in complete agreement as to the direction (U.S. Bureau of Reclamation 2011). Many projections indicated decreasing precipitation, so less certainty is associated with these trends than is the case for temperature (U.S. Bureau of Reclamation 2011).

Changes in temperature and precipitation will both affect snow accumulation during the late autumn through early spring, however, it is projected warming that seems to dominate projected snowpack changes (U.S. Bureau of Reclamation 2011). Warming is expected to decrease snow accumulation, with decreases being more substantial in areas, such as the eastern Plains, that have cool season temperatures closer to freezing thresholds (U.S. Bureau of Reclamation 2011). The ensemble medians indicate decreases in snow-water equivalent, for the basin as a whole, on April 1 of 76%, 81%, and 84% for the 2020s, 2050s, and 2070s, respectively, as compared to the 1990s (U.S. Bureau of Reclamation 2011).

General Circulation Model projections of future climate through 2099 indicate a wide range of possible scenarios (IPCC 2007). To determine the sensitivity and potential effect of long-term climate change on the freshwater resources of the United States, Markstrom et al. (2011) selected fourteen basins from across the United States and modeled them with the Precipitation Runoff Modeling System surface water hydrology model (Markstrom et al. 2008). Two of these fourteen basins fall within the Great Plains Regional Assessment area. The Starkweather Coulee Basin, North Dakota (Vining 2002) and the South Fork of the Flathead River Basin, Montana (Chase 2011) were both the subjects of previous Precipitation Runoff Modeling System modeling studies.

The Starkweather Coulee Basin exhibits little to no stream flow from September through February, mainly because of the sub-freezing temperature in the basin. This is not projected to change substantially. As projected temperatures increase, evapotranspiration increases, resulting in less stream flow available for runoff and storage. In the South Fork of the Flathead River Basin, seasonal stream flow is projected to increase from November through April and decrease in May, June, and July by the end of the 21st

Table 4.2 Great Plains Water Shortage Risk and Crop Value in At-Risk Counties, by State

State	Percent of Counties At-Risk	Total At-Risk	Extreme Risk	High Risk	Moderate Risk	Value of Crops Produced in At-Risk Counties (in $1,000s)
Colorado	55%	35	12	15	8	$1 ,484,453
Kansas	86%	90	41	20	29	$4 ,197,856
Montana	46%	26	1	17	8	$7 37,187
Nebraska	97%	90	46	27	22	$6 ,423,909
New Mexico	82%	27	10	9	8	$3 50,376
North Dakota	83%	44	0	4	40	$ 3,895,935
Oklahoma	91%	70	25	27	18	$ 8 91,167
South Dakota	56%	37	0	7	30	$ 1,863,979

Source: (Roy et al., 2010)

century. These changes correspond to changes in mean monthly snowmelt (Markstrom et al. 2011).

Numerous sources of uncertainty have been identified in this study. Large uncertainties are associated with the representation of the physical processes, model structure, and feedbacks within the climate system as projected by the global climate models. The scenarios chosen for this study represent different economic, social, political, and technological development for the future, none of which may be the actual path (Hay et al. 2011).

To date, there have been multiple studies that have used climate models to try to predict future water availability in the Great Plains (Table 4.2). But these studies tend to have significant uncertainties on regional or watershed scales, and they often come up with varying results depending on the methodology, climate and hydrological models used, downscaling techniques, and the assumptions that go into the models (Thomson et al. 2005, Mehta et al. 2011). These uncertainties call for close and continued partnership between climate researchers and resource managers, ideally using iterative, risk-based approaches that can be flexible and incorporate a range of scenarios into planning (Pulwarty 2003, Vogel and O'Brien 2006, Kallis 2008, Brekke et al. 2009, Huntjens et al. 2010, May and Plummer 2011). These climate impacts on water resources will have consequences associated with energy generation and operations throughout the Great Plains (as will be discussed in later chapters). In addition, extraction of natural resources associated with energy development in the region will also be affected.

Water Infrastructure

In the semi-arid region of the Great Plains, an extensive system of water-related infrastructure has been developed to provide for a more stable water supply for agriculture. In addition, this infrastructure also provides flood control, hydroelectric power, navigation, recreation, stormwater management, and wastewater treatment. Components include dams, reservoirs, pipelines, irrigation canals, wells, pumps, water treatment systems, dikes, levees, floodgates, hydroelectric plants, storm sewers, wastewater treatment systems, and more (Western States Water Council 2011). This system operates within a variable precipitation regime due to seasonal patterns of rain and snow fall.

Some of the important water infrastructure components for the major Great Plains river basins are discussed below. This is followed by a discussion of the American Society of Civil Engineers (ASCE) national report card on infrastructure and the current status of dam and drinking water infrastructure in the Great Plains. The chapter ends with some notes on infrastructure recommendations made by the Western States Water Council, which is comprised of representatives appointed by the governors of 18 Western states, including all of the states in the Great Plains region.

In the Missouri River Basin, water infrastructure development intensified after 1902 with the passage of the Reclamation Act, which established irrigation support in the western US. In 1937, the first of the Missouri River's main-stem dams was constructed at Fort Peck, MT, as part of a Works Progress Administration project, to provide minimum flows for downstream navigation. (National Research Council 2002, Benke and Cushing 2005). In 1944, the US Army Corps of Engineers and Bureau of Reclamation basin management plans for the Missouri River Basin were merged in an agreement known as the Pick-Sloan Missouri Basin Program (National Research Council 2002, U.S. Bureau of Reclamation 2011). The stated goals included providing flood control, irrigation, navigation, power, water supply, wildlife, and recreation (National Research Council 2002, U.S. Bureau of Reclamation 2011). The Pick-Sloan program resulted in the construction of five main-stem dams downstream of Fort Peck and over 40 dams on basin tributaries.

Today, owing to a variety of projects, the Missouri River Basin contains over 17,200 reservoirs, providing a storage capacity of about 141 million acre-feet (174 billion m^3), 73.4 million acre-feet (91 billion m^3) of which are provided by reservoirs behind six US Army Corps of Engineers-built main-stem dams. It is the largest reservoir system in the US (U.S. Army Corps of Engineers 2006). Three of the main-stem reservoir lakes (Fort Peck in Montana, Sakakawea in North Dakota, and Oahe in South Dakota) are among the largest human-made lakes in the country, behind only Lakes Mead and Powell (National Research Council 2002). The combined surface area of the six US Army Corps of Engineers main-stem reservoirs at normal pool levels is about 1 million acres (404,700 hectares), and the reservoirs provide fish and wildlife habitat as well as recreational opportunities. However, the large surface area also leads to considerable evaporation losses, which vary from year to year, but are estimated to average about two million acre-feet per year (2467 million cubic meters per year) (U.S. Army Corps of Engineers 1998).

In addition to providing water for irrigation, domestic, municipal, and industrial uses, Missouri River water plays a key role in electricity generation for the region. Twenty-five thermal-electric power plants along the main-stem river use either reservoir or river water for cooling and, together, have a gross generation capacity of about 14.2 million BTUs/second (15,000 MW) (Benke and Cushing 2005, U.S. Army Corps of Engineers 2006). Hydropower from six of the main-stem dams (Fort Peck and others further downstream) contributes an additional 2.3 million BTUs/second (2435 MW) of capacity (Benke and Cushing 2005).

In the Arkansas River Basin, one of the main water infrastructure developments is the McClellan-Kerr Arkansas River Navigation System, which is an extensive series of locks and dams on the White, Arkansas, and Verdigris Rivers that ensure that barge traffic can move year-round between the Tulsa, Oklahoma Port of Catoosa and the Mississippi River (U.S. Army Corps of Engineers 2012). The 445-mile (716 km) long system includes a 377-mile (607 km) stretch of the Arkansas River and a 9-mile (14.5 km), Arkansas Post Canal that connects the White and Arkansas Rivers (U.S. Army Corps of Engineers 2012). Resources shipped through the McClellan-Kerr system include agricultural products, petroleum, and coal (Encyclopedia Britannica Online 2012). In addition to enabling navigation, the system, dedicated in 1971, provides flood control and hydroelectric power (Encyclopedia Britannica Online 2012).

In the Red River of the South Basin, Lake Texoma is an important reservoir, located at the junction of the Red and Washita Rivers. Lake Texoma is the twelfth largest lake in the US, in terms of capacity, and serves a variety of purposes. It is one of the few reservoirs in the US in which striped bass reproduce naturally (U.S. Army Corps of Engineers 2010), and is used extensively as a recreational facility. Since the opening of Denison Dam in 1944, the project has prevented cumulative flood damages of over $852 million in 2008 average dollars (U.S. Army Corps of Engineers 2010). Hydropower facilities have a generating capacity of 66,347 BTUs/second (70 MW). In terms of water supply, the lake currently has one full-time user, the Greater Texoma Utility Authority, acting for the City of Sherman, Texas. However, anticipated regional population growth has caused both Texas and Oklahoma to start securing rights to the lake water for future water usage. Potential future customers include the Dallas-Fort Worth metropolitan area (Benke and Cushing 2005). Lake Texoma is also a popular recreational facility with over six million people visiting annually. In addition, the lake provides two state parks, two national wildlife refuges, and is one of the few reservoirs in the US in which striped bass reproduce naturally (U.S. Army Corps of Engineers 2010).

In the Texas Gulf hydrologic unit, important reservoirs include Lake Livingston in the Trinity River Basin and Lakes Conroe and Houston in the San Jacinto River Basin, all of which supply surface water for Houston (City of Houston 2012). Lake Livingston accounts for 75% of Houston's surface water supplies (Trinity River Authority of Texas 2010). Lakes Lewisville, Grapevine, Ray Roberts, and Ray Hubbard in the Trinity River Basin all supply water for Dallas, as does Lake Tawakoni in the Sabine River Basin

(Trinity River Authority of Texas 2010, Dallas 2012). Lake Fork in the Sabine River Basin and Lake Palestine in the Neches River Basin are on reserve for future Dallas water supply (Dallas 2012). In the Rio Grande River Basin, some of the main reservoirs are the Cochiti, Elephant Butte, and Caballo in New Mexico, and the Amistad and Falcon, both of which are international, shared by Texas and Mexico.

Infrastructure Status

A 2009 ASCE Report Card gives the nation's dam infrastructure a grade of "D" and the nation's drinking water, wastewater, levees, and inland waterways infrastructure grades of "D-" (American Society of Civil Engineers 2009, Western States Water Council 2011). The status of dam and drinking water infrastructure in the Great Plains is discussed in more detail below.

As evidenced by the river basin infrastructure descriptions, dams abound throughout the Great Plains and, thus, dam safety is a concern. Dams may be considered deficient because of aging and deterioration, lack of maintenance, or because of increased engineering knowledge about the ability of a dam to withstand large flood events or earthquakes. According to a 2009 ASCE report card on the nation's infrastructure, the two states in the Great Plains region that had the highest number of dams in need of rehabilitation to meet applicable dam-safety standards were Oklahoma and South Dakota with 150 and 67 dams, respectively. Dams in the high hazard category are those that if they fail are anticipated to result in loss of life. According to the ASCE report, over 85, 40, 25, and 20 high hazard dams in Texas, Wyoming, Oklahoma, and South Dakota, respectively, had no Emergency Action Plan (EAPs) as of 2008 (American Society of Civil Engineers 2009). ASCE recommended that all high hazard dams throughout the US develop EAPs by 2011.

In the context of climate change, planners will need to factor in new levels of safety that take changing peak flows and precipitation regimes into account in dam design, operation, and regulation (State of California Department of Water Resources 2008). In addition, more extreme rainfall events may increase soil erosion and bank failure, which could increase sedimentation behind dams.

In 2007, the US Environmental Protection Agency (EPA) conducted its fourth survey and assessment of the nation's drinking water infrastructure needs, studying, in particular, the twenty-year (2007-2026) capital improvement needs for water systems to continue providing safe public drinking water. The results of the survey noted that much of the nation's drinking water infrastructure is approaching or has already reached the end of its design life and is now in a "rehabilitation and replacement" stage (U.S. Environmental Protection Agency 2009). This is reflected in the over $320 billion estimate of infrastructure investments needed by the nation's drinking water utilities over the next 20 years (in average January 2007 dollars).

The 2007 survey identified an emerging need for new source water infrastructure required to address existing or anticipated drought conditions. In its 2011 survey, the EPA included supplemental questions related to climate readiness (U.S. Environmental Protection Agency 2011), but results of that survey are not yet available.

Western States Water Council Recommendations

The Western States Water Council (WSWC) is an advisory group that reports to the Western Governors Association and is tasked with helping to ensure that Western states have an adequate, sustainable water supply now and in the future. A 2011 report documents recommendations made by individual participants attending a workshop on Western Water Resources Infrastructure Strategies: Identifying, Prioritizing, and Financing Needs. Among the many recommendations were ones related to emphasizing water conservation as a crucial strategy that can delay or reduce the need for developing new water supplies and related infrastructure. Participants identified tools that promote water conservation, including full-cost pricing strategies that account for water scarcity, approaches that reduce per-capita demand, and programs to monitor and address leakage. Another group of recommendations revolved around the diversification of local water supply sources; for instance, through water reuse, the use of brackish groundwater, and desalination. Lower quality water (e.g. brackish groundwater, reclaimed wastewater) could be used for nonpotable purposes, while higher quality water could be reserved for potable uses. Investment in green infrastructure was also proposed as a cost-effective approach to managing stormwater and conserving water. In terms of financing, the WSWC report noted that public-private partnerships are one option that could make it easier to finance water infrastructure projects. The WSWC report also recommended that state and federal agencies examine their ability to provide assistance to small communities, many of which are located in rural areas and many of which lack the resources to finance needed projects. The report notes that although local water supplies should be developed first, interbasin water transfers and markets are options that may be necessary.

Managing Water in the Great Plains

The availability of water is critical to the viability and prosperity of the Great Plains region. Water scarcity -- through both the legal over-allocation of existing water resources and a relative decrease in physical availability from climate change -- is quickly becoming one of the greatest challenges in the Western United States. In the Great Plains, the trend is for people and water to move from rural areas to cities. To accommodate increasing population growth and development in certain areas and rising energy needs, water is increasingly going to be reallocated to "higher valued uses" (Western Governors' Association 2006). A challenge to the region is develop policies and management frameworks that are flexible and responsive to the variability of water resources and demands. These regulating instruments need to address vulnerabilities of local communities and ecosystems under uncertainties of climate and other social-ecological dynamics occur across the Great Plains region.

There are many complex legal and policy issues when it comes to water allocation and this can sometimes lead to conflict (Bell and Taylor 2008). One example is in the Pumpkin Creek watershed in Nebraska where surface water irrigators have taken groundwater irrigators to court, claiming that their excessive groundwater use prevents the surface irrigators from being able to withdraw their appropriations (Knutson 2008). As climate change impacts streamflows and water availability, further complications and these types of legal battles can be anticipated.

In snowmelt dominated river basin systems (i.e., the western Great Plains), the possibility of climate change shifting seasons to result in earlier timing of runoff has implications for water use and management in states where there are timing regulations built into water rights. Examples include where state laws specify when certain users can divert and use water. Earlier snowmelt and runoff could lead to user impacts, management problems, and legal conflicts if runoff timing is mismatched with the irrigation season (Kenney et al. 2008). To date, this has not resulted in litigation, but water managers and irrigators are increasingly concerned about the implications of this issue. States in the Great Plains vary in terms of whether their water rights systems have explicit timing requirements, however, multiple interstate compacts in the Great Plains have timing requirements, which could result in additional legal conflicts between states over water rights and allocations (Kenney et al. 2008).

Groundwater rights and management are also defined by each state. In the High Plains Aquifer region, the three states that overlay most of the aquifer and withdraw significant amounts of water for irrigation of agriculture are Kansas, Nebraska, and Texas. All three states use different doctrines for groundwater allocation (Peck 2007). This means they have very different laws and institutions for managing and regulating water, so collective efforts for managing impacts and adaptations will require interstate cooperation well beyond anything already experienced to date, particularly with the implications of diminishing shared aquifer water and changes in hydrological cycles, water availability, and recharge rates. Such efforts will require increased cooperation and mechanisms for conflict resolution between states, local users, and the federal government (Peck 2007), and approaches that incorporate the understanding of climate variability and change into institutional knowledge and decision- and policy-making. These policies and water management arrangements will be more difficult as climate change impacts affect water availability across the Great Plains.

In some regions of the Great Plains, the establishment of water rights is still unsettled and the process of quantifying water rights can be time-consuming, costly, and complex (Colby et al. 2005). It can sometimes take decades to complete and involves a number of specialists to determine water allocation, water rights, which crops are sustainable, and how much water is needed to grow them, as well as other issues. Negotiations can cover issues beyond the settlement of priority dates and the quantification of water rights, however. These policy instruments can provide additional flexibility for addressing deficiencies in state and federal policies -- for instance, with respect to hydrologic connections between groundwater pumping and streamflows -- and allow for more integrated water resource management from both stakeholder and environmental points of view (Colby et al. 2005). Water can be reallocated or new development projects agreed upon. Many tribes may not have the financial capacity to convert paper water rights entitlements to actual wet water infrastructure, and sometimes provisions in negotiated settlements can include financial backing. Given the expected increase in climate variability, one particularly important aspect of negotiations, could be the agreement among users regarding water allocation and management during wet versus dry years. Another important aspect could be agreement on the selling or leasing of water and on subsequent profit-sharing.

Although the physical engineering in the Great Plains region has provided benefits, there have also been some costs. In particular, riverside Native American communities were relocated when their fertile floodplain homelands were inundated as reservoirs were created. In addition, the engineering has greatly reduced the amount of sediment transported by the Missouri, which has altered riverine habitat important to some native biota. This has contributed to the listing of two bird species, the least tern and the piping plover, and one fish species, the pallid sturgeon, under the federal Endangered Species Act (National Research Council 2011). The changes in the Missouri's sediment transport regime have also resulted in channel bed lowering, which is causing problems for infrastructure by eroding bridge foundations at many sites, foundations of flood protection structures in and near Kansas City, and lowering water levels at municipal intakes (National Research Council 2011). Similar issues are of concern on other river systems where built water infrastructure has had unintended consequences.

The socio-economic dynamics of the Great Plains region, such as increased urbanization, population loss in rural counties, aging infrastructure, and loss of social services have increased vulnerability to water stress regardless of climate change. Additionally projected climate change indicates increased drought risks and impacts and the need to enhance drought preparedness measures (Knutson 2008). Since the epic drought of the 1930s, many programs and adaptations have been put in place to buffer risk from drought. Despite these measures, considerable vulnerability to water stress and drought still exists because of the continual expansion of and competition between water users, changing water availability, and various management strategies that have had unintended consequences or varying impacts on different stakeholders (Knutson 2008).

Pumping of the Ogallala Aquifer has lowered the water table so much in certain areas that in Nebraska they have begun to issue moratoriums on new well drilling in several basins (Knutson 2008). Conservation has been implemented in some areas throughout the region, however, a tradeoff is that this leads to reduced return flows for downstream users (Knutson 2008). Interviews with agricultural producers in Nebraska found that they identified lack of capital and the need to respond to markets as barriers to adapt to drought risks (Knutson and Haigh 2011).

McLeman and colleagues (2008) used analog studies to look at past responses to drought vulnerability, to identify lessons learned that can help place climate change adaptation within the context of overall vulnerabilities and adaptive capacities to a broad range of socio-economic conditions. Studies on the 1930s drought in Oklahoma highlighted that different demographics had different adaptive capacities and therefore adapted differently (e.g., land owners versus tenant farmers); the important role that social capital and social services played in sustaining livelihoods; and the critical role that federal programs played such as the Agricultural Adjustment ActWorks Progress Administration and the Farm Security Administration (McLeman and Smit 2006).

While social and environmental conditions have changed since the 1930's, these lessons can provide valuable insights into understanding how communities and governments respond to drought.

While significant water resources vulnerabilities exist from climate change, a new paradigm for water policy and management is emerging. The top-down water planning of the past is being replaced by new and innovative solutions through local stakeholder processes that incorporate the needs of communities into state planning (Western Governors' Association 2006).

Chapter 5

Ecosystem and Biodiversity Conservation Issues

Climate-ecosystem interactions and the inherent uncertainty associated with a variable and changing climate pose a formidable threat to the region's biological diversity and the function of aquatic and terrestrial ecosystems. Recent alterations of seasonal trends and extreme events (i.e., droughts, heat waves, floods, etc.) have affected ecosystem functions and triggered thresholds of physiological and life-cycle patterns of various species. These changes have led to changes in habitat conditions and species composition shifts. These threshold changes also have impacts on species mortality and the persistence of plant and animal populations (Allen 2010). The invasion of exotic species into terrestrial systems is likely to accelerate in response to longer growing seasons, because they will have more time to establish themselves.

Climate change projections for the Great Plains present a diversity of possible stressors which would exacerbate the current environmental challenges of many wildlife and conservation management efforts. Extreme events such as droughts, floods, heat waves, and winter storms result in alterations of plant communities, changes in hydrological dynamics (e.g., stream flow and wetland innundation), and provision of ecosystem services (U.S. Climate Change Science Program 2008), which can result in threshold changes of critical ecosystem level functions across the region (U.S. Climate Change Science Program 2009).

Aquatic systems, in particular, are being pushed to their limits due to habitat destruction and warming water in which many species could experience temperatures beyond their thermal tolerances (Covich et al. 1997, U.S. Climate Change Science Program 2009). Rising temperatures and increasing water demands will stress aquatic systems beyond sustainable capacities. Impoundments and diversions have a major impact on the hydrological flow. While many aquatic organisms have made adaptive changes to these flow modifications, additional climate changes can increase the vulnerability of organisms to additional changes in water flow during critical breeding periods or migratory timing of key species across the region.

Hydrological regime changes and water temperatures can affect various species differently across the region. Formerly perennial streams are now observed to flow intermittently resulting in changing plant and animal populations residing in these streams, ponds, and wetlands. Warmer water temperatures decrease oxygen retention, thereby increasing stress on many aquatic organisms. Simultaneously, an aquatic species' oxygen demand will be elevated as metabolic rates increase in response to warmer water.

Understanding the rate of change in temperature and precipitation will likely be as important as understanding the long-term endpoint. Natural systems in the Great Plains have evolved with high levels of climatic variability and have many built-in mechanisms that allow them to be somewhat resilient to climate change. Such resiliency, however, depends on sufficient time for adaptation. If climate change occurs rapidly, natural systems may not be able to adapt at a rate that ensures their survival – leading to a loss in regional biodiversity and local extinctions.

Climate Change and Fragmentation

The highly productive Great Plains ecosystem originally consisted of about 500-million acres or 2 million km^2 of intact grasslands that supported huge herds of bison and other ungulates. Major evolutionary grassland drivers were climatic variation, herbivory by nomadic ungulates, fire, and, in the central and western grasslands, prairie dogs (Axelrod 1985, Anderson 2006). Populations of all native grassland ungulates and prairie dogs have been hugely reduced, and fire regimes no longer mimic natural processes. Furthermore, both terrestrial and aquatic habitats are extensively fragmented due to agriculture, roads, energy infrastructure, and water impoundments, with consequent effects on biota and ecological processes (The Heinz Center 2008, Sabo et al. 2010).

The loss of biodiversity in the Great Plains has been driven by habitat loss, degradation, and increasing fragmentation, with future biodiversity also subject to changes as a result of climate change (Becker et al. 2007). The combined effects of climate change and land use change are key threats to ecosystem processes and biodiversity in the Great Plains. Many species are responding to rising temperatures by shifting distributions, apparently at increasingly greater rates (Parmesan 2006, Chen et al. 2011). The simultaneous loss and fragmentation of habitats impedes the ability of species to move into new areas in response to rapid climate changes. In the Great Plains, the extensive network of roads and agriculture has resulted in highly fragmented grasslands – more than 85% of all intact grassland patches are now less than 100 mi^2 (260 km^2) in area (The Heinz Center 2008). Connectivity within the landscape is considered a foundation for preserving biodiversity in the face of climate change (e.g. Kostyack et al. 2011).

In the southern Great Plains, habitat loss, degradation, and fragmentation are due overwhelmingly to land conversion for agriculture, with over 70% of the land surface altered, and over 90% in some areas (Gray et al. 2004). Over 70% of playas >4 ha in basin area in the southern Great Plains have been modified for agriculture (either tilled or excavated with pits to gather irrigation return water) (Guthery and Bryant 1982). In 1965, only about ~0.6% of playas in Texas were modified; by 1981, ~43% were (Nelson et al. 1983), so these changes are recent and severe. Land conversion alters wetlands and their biota by changing water chemistry, hydroperiod, and sheer presence of wetlands themselves. Land conversion to agriculture has been shown to greatly increase sedimentation within playas surrounded by cropland relative to indigenous grassland (Luo et al. 1997, Tsai et al. 2007), and sedimentation is considered the primary threat to playas (Smith et al. 2011). Playas within a tilled watershed typically experience a shorter hydroperiod relative to playas in untilled watersheds. The mechanism is unclear, but is possibly the result of reduction in basin volume as sediment depth increases, thereby inducing volume overflow and increased evaporative loss (Luo et al. 1997, Tsai et al. 2010), or from sediments keeping hydric soil cracks open and thereby facilitating infiltration (Ganesan 2010). Playas surrounded by cropland contain 8.5 times more sediment than playas in more natural situations (Luo et al. 1997), which buries seed and insect egg banks (Gleason et al. 2003). Landscape fragmentation in the southern Great Plains has also been shown to impede the overland dispersal of amphibians (Gray et al. 2004), thereby effectively isolating wetlands. Generally speaking (not just Great Plains), species richness is lower in isolated wetlands for various insects (e.g. notonectids and dytiscids) (Wilcox 2001, McCauley 2006).

Phenology

Changes in the timing of seasonal or phenological events—such as flowering, migrations, and breeding—have been called a 'globally coherent fingerprint of climate change impacts' on plants and animals (Parmesan 2007). Phenological shifts can result in perverse ecological effects, as there is a desynchronization between migratory birds and their prey, or pollinators and flowers. Simple shifts in phenology, as described below, can serve as sensitive and integrative indicators of climate change. More complex interactions between species and ecological processes are more difficult to detect, partly because baseline data are sparse.

Climate-induced changes in phenology have been linked to shifts in the timing of allergy seasons and cultural festivals, increases in wildfire activity and pest outbreaks, shifts in species distributions, declines in the abundance of native species, the spread of invasive species, and changes in carbon cycling in forests. Scientists have documented these fingerprints of climate change across the US using a variety of data sources. From Texas to Saskatchewan, the length of pollen season for ragweed (Ambrosia spp.), a common human allergen, has increased from 1995-2009 by as much as 16 days in certain areas (e.g., Fargo, North Dakota). This increase was correlated with an increase in frost-free days as well as later onset of first frost in the fall, but not with annual precipitation (Ziska et al. 2011). The mean egg-laying date of American Pipits (Anthus rubescens) has become approximately 5 days earlier, and mean clutch size has increased by 0.2 eggs in the mountains of Wyoming from 1961-2002. These changes were significantly related to earlier snowmelt, which occurred about 7 days earlier (Hendricks 2003). Using data from six locations throughout the Great Plains, it was observed that winter wheat is blooming 6 – 10 days earlier now than it was 70 years ago. Warming spring temperatures have also been observed over this same period (Hu et al. 2005).

PREDICTING FUTURE COMMUNITY COMPOSITION AND TIMING

Based on how closely Konza prairie plant community flowering tracks environmental conditions (Craine et al. 2011) and its predictable differences from other floras, informed predictions about how future climate change may alter plant communities are possible. In Konza, where regional climate models consistently predict warmer future

BOX 5.1

Case Study: Shifts in Flowering Phenology in the Northern Great Plains Over 100 years

First flowering dates were compared for 178 species of plants from 1910–1961 and 2007–2010 in North Dakota. During this time period, temperatures increased 3°F (1.7 °C) from the first 9 years of the study to the last 9 years, and growing season duration increased from 132 days to 154 days. Between 24% and 41% of plants showed a change in flowering time over the period, and even more species showed an earlier first flowering date in the two warmer years of this study (2007, 2010). It would be expected that these species that showed a strong response to climatic variables will show a continued response with increasing temperatures. More than 50% of the species did not show a change from earlier in the century. The reasons for this are unclear, but it is possible that the phenology of these species is not as tied to temperature or precipitation (Dunnell & Travers, 2011).

temperatures along with a more variable precipitation regime (Christensen et al. 2007), a series of growing-season shifts may alter flowering. First, as found with many floras globally, early-season species may shift earlier as thermal sums required to trigger flowering are met earlier. Additionally, species invasions from donor floras may increase (Wolkovich and Cleland 2011) as the Konza season expands to increase overlap in phenological climatic space with plant communities such as those located in Europe. Additionally, as the mid-growing season drought may become more pronounced, the number of species flowering mid-season may be reduced. Evidence of such a shift towards a novel mid-season gap (or decrease) in flowering has already been suggested in other floras observationally (Aldridge et al. 2011) and via experiments (Sherry et al. 2007). Comparing the responses of flowering phenology to experimental warming and the differences in flowering between Konza in Kansas and Fargo, North Dakota suggest a common inflection point around which flowering changes with changes in temperature. In an Oklahoma grassland experiment, warming caused early-flowering species to flower earlier and late-flowering species to flower later with an inflection point near mid- to late-July. This date range is similar to the July 14 inflection point for changes in flowering dates between Konza and Fargo. The universality of this mid-July date remains to be seen, but it appears to serve as a consistent benchmark for predicting the responses of flowering phenology to warming (Sherry et al. 2007).

PHENOLOGICAL INDICATORS – EXTENDED SPRING INDEX

Schwartz et al. (2006) provided a set of modeled and derived pheno-climatological measures that reflect increasing temperature in the northern hemisphere. Schwartz et al.

(2006) developed their spring index from station-level weather observations and confirmed the efficacy of the index from observations of cloned lilac and honeysuckle. The spring index has now been extended to areas outside the range of lilacs and honeysuckle (McCabe et al. 2011) and it reveals that first leafing and blooming dates have increased by as much as eight days since the 1950s in the northern Great Plains

Effects of Climate Change on Vegetation and Ecosystems of the Great Plains

Land cover and land use across the Great Plains is dominated by livestock-based agriculture, especially cattle and croplands. However, within this matrix, untilled remnants of natural prairie retain ecosystems and habitats of the High Plains region as an interspersed network of managed rangelands and natural areas. Agriculture has typically reduced the nutrient capacity of Great Plains soils through tillage and biomass extraction (Peterson and Cole 1995). However, ungulates and grazing animals typically develop a somewhat symbiotic relationship with productivity patterns and nutrient cycling (Augustine et al. 1998), suggesting that natural patterns may be retained with some agro-economic systems.

In general, patterns and dynamics of Great Plains grassland ecosystems are driven by climate and soil patterns with additional influences on species composition, biomass production and nutrient cycling induced by herbivory (livestock, wildlife, and insects) and biological responses, differences in plant nutrient use efficiency, water use efficiency, wildfire, plant disease, nutrient cycling and biomass decomposition. All of these are potentially affected, directly and/or indirectly by climate change (e.g. King et al. 2004, Morgan et al. 2008). Large rainfall events, especially after periods of prolonged drought, interact with exposed and tilled soils generating significant quantities of sedimentation and topsoil degradation (LaGrange et al. 2011).

Strong gradients of temperature and precipitation help define the composition and structure of vegetation across the Great Plains (Lauenroth and Burke 1995, Peterson and Cole 1995). Mean annual temperatures increase from 39 °F (4°C) in Montana to 68 °F (20°C) in central Texas. Generally, the optimum temperature for photosynthetic rate in C_4 plants is higher than that for C3 plants (Black 1973, Ehleringer and Bjorkman 1977, Epstein 1998), however, experimental trials at Long-Term Ecological Research sites indicate that increased concentrations of CO_2 decrease actual evapotranspiration and increase efficiency of gas exchange, disproportionately favoring C_4 species.

Temperatures are projected to continue to increase across the Great Plains over this century, with summer changes projected to be larger than those in winter especially in the south-central plains (Christensen et al. 2007). The average temperature in the Great Plains already has increased roughly 1.5°F (0.8 °C) relative to a 1960s and 1970s baseline. By the end of the century, temperatures are projected to continue to increase somewhere between 2.5°F (1.4 °C) and more than 13°F (7 °C) compared with the 1960 to 1979 baseline, depending on future anthropogenic emissions. Specific ecosystem effects of warming are unclear, given the complexities of interactions with soils, nutrients, CO_2, grazing and fire. Warming experiments in tallgrass prairie suggested increasing soil temperatures 3.5 °F (2°C) extended the growing season and yielded greater aboveground productivity, but did not affect belowground productivity (Wan et al. 2005). Whereas, in a mixed grass ecosystem, warming the canopy 5.5 °F (3°C) increased nitrogen use (Dijkstra et al. 2010) without clear, overall effects on above- or belowground productivity (Morgan et al. 2011). Projected increases in temperature, evaporation, and drought frequency add to concerns about the region's declining water resources. Water is the most important factor affecting activities on the Great Plains.

Changes in temperature affect the rates of chemical reactions and the exchanges of energy between the land and the atmosphere. These temperature effects on biological responses have the potential to increase plant growth (Luo et al. 2009a), speed up plant development (Cleland et al. 2006, Sherry et al. 2007, Hovenden et al. 2008b), and increase the decomposition of soil organic matter (Rice et al. 1998). These same potential effects can also be limited by soil moisture. As a result, warming may increase the plant growth in rangeland systems in years with adequate moisture, but have little or even negative effects when soil moisture is inadequate and warming leads to increased evapotranspiration rates and desiccation (Xia et al. 2009, Pendall et al. 2010, Fay et al. 2011, Morgan et al. 2011).

Vegetation responses to rising atmospheric CO_2 concentration, warming, and other climate changes are regulated by interactions with independent variables, including soil type, which strongly influences plant and soil water relations; the regional species pool from which new species may enter an ecosystem; the disturbance regime; and synoptic climate. The disturbance regime and available species pool at any given location may be decisive in dictating vegetation responses to climate change. In general, however, each of the primary climate change drivers, including CO_2 enrichment, warming, and an anticipated increase in precipitation variability and extreme weather events, influence vegetation by affecting soil water availability to plants. Given the strong imprint of the east-to-west gradient of declining precipitation on the composition and structure of semi-natural vegetation in the Great Plains, we anticipate that the collective effect of climate change drivers on vegetation will be manifested mainly through changes in soil water availability. These effects are evident in manipulative experiments with each of the individual aspects of climate change. For example, CO_2 effects on vegetation composition usually are linked to the water-savings effects of CO_2 enrichment on grasslands (Morgan et al. 2004a).

CO_2 enrichment has modified species abundance in ecosystems as diverse as Swiss grasslands and semi-arid shortgrass steppe by slowing soil water depletion and preferentially increasing seedling recruitment of certain species (Morgan et al. 2004b, Niklaus et al. 2004). In contrast, CO_2 had little effect on species abundances in C_4-dominated tallgrass prairie in Kansas (Owensby et al. 1999), presumably because the growth of the shorter C_3 species was limited by low light or nitrogen availability, or C_3 plants were incapable of exploiting the mid- to late-season improvement in soil water that occurred at elevated CO_2.

Tallgrass prairie has been reduced to 1% of its historic land cover in North America (Samson and Knopf 1994) and, unfortunately, what remains of the Great Plains is being threatened by global change factors in addition to climate. Continued land-use change, woody vegetation encroachment, plant invasions, and anthropogenic increases in nitrogen are of high conservation concern in the Great Plains region. Individually, these global change factors have serious consequences for community composition and ecosystem function, and each of these drivers has the potential to interact both directly and indirectly with climate change. Land-use change through conversion of native grasslands into cultivated cropland results in decreased soil carbon storage, decreased biodiversity, and increased soil erosion (Davidson and Ackerman 1993, Parton et al. 2005). Changes in grassland management through grazing and fire regimes have strong impacts on ecosystem health. Typical domestic grazing practices and fire suppression on Great Plains grasslands cause declines in species diversity (Leach and Givnish 1996, Collins et al. 1998) and negatively impact ecosystem function.

Therefore, untilled rangelands offer the most promising reserve of native species and functioning Great Plains ecosystems. Meanwhile, the fragmented distribution of these lands represents the framework for a spatial distribution of native plants and animals across the region in the future. But restoration of landscape-scale processes, especially in the context of climate change, presents a critical challenge for managers, planners and society.

ADDITIONAL STRESSORS

The removal of grazing and the suppression of fire from the Great Plains cause a decline in species diversity (Leach and Givnish 1996, Collins et al. 1998) and negatively impact ecosystem function. Fire suppression has caused an increase in woody plant encroachment (Bragg and Hulbert 1976, Schmidt and Leatherberry 1995) in the Great Plains. *Juniperus virginiana* and *Cornus drumundii* are two woody plant species of conservation concern in tallgrass prairie because of encroachment into native prairies modifying productivity patterns (Norris et al. 2001b, Lett et al. 2004, Lett and Knapp 2005) and decomposition dynamics (Norris et al. 2001a), which has consequences for regional carbon storage. In addition to woody encroachment, non-native plant species are invading the Great Plains, many of which are C_3, cool-season annual grasses (Cully et al. 2003). Extreme climatic events may increase plant invasions since disturbance is positively associated with such patterns (Hobbs and Huenneke 1992). In addition, both native and non-native species have the potential to become invasive as grazing and fire regimes are altered (Simberloff 2008), as climate and humans expand the potential habitat of species (Barney and DiTomaso 1977), and as monocultures of crops increase in land cover (Hartman et al. 2011). Complicating matters, woody plants have the potential to survive, and even thrive, with altered precipitation patterns as they access water from deeper soils than the dominant prairie plant species do (Ratajczak et al. 2011).

Significant amounts of atmospheric nitrogen deposition, primarily from burning of fossil fuels, continue to be deposited across regions that are typically nitrogen limited. Increased nitrogen inputs due to both atmospheric deposition and runoff from agricultural areas (Vitousek et al. 1997) will likely continue to have large effects on the plant communities of the Great Plains. Nitrogen has a stronger effect on plant communities where water is not the primary limiting factor, such as in mixed and tallgrass prairies. Increased nitrogen availability tends to result in decreased plant diversity, while increasing plant production (Wedin and Tilman 1993, Gough et al. 2000, Clark 2007). Forbs and woody plants have been shown to increase in abundance with nitrogen, with the dominant C_4 grasses decreasing in abundance (Seastedt et al. 1991, Briggs et al. 2005, Bond 2008).

The turnover in plant community composition as a result of global change factors may have strong functional consequences for the way prairie systems respond to altered precipitation and temperature patterns. The current prairie community, dominated by perennial C_4 grasses, is well adapted to deal with high variability in rainfall and temperature (Knapp and Smith 2001, Weltzin et al. 2003, Huxman et al. 2004). However, the decline of these dominant grasses due to one or several of the potential mechanisms would have unknown, but likely detrimental, consequences. For example, due to efficiencies of the C_4 pathway (versus C_3 pathway), the newly formed communities may be poorly adapted to precipitation and temperature variations. Considering the phenology and functional traits of species that dominate these altered communities will prove important for estimating local climate change effects on the Great Plains prairie systems into the future.

Freshwater Ecosystems

DEPRESSIONAL WETLANDS

Two main types of wetlands in the Great Plains form a collective network of aquatic habitat in an otherwise semi-arid region. In the northern Great Plains, prairie pothole wetlands are glacially formed and heterogeneous in structure and hydrology. In the central and southern Great Plains, playa wetlands are aeolian equivalents of prairie potholes, but are far more uniform in shape and structure. Both types of wetlands are runoff-fed with variable hydroperiods that range from temporary to effectively permanent.

Pothole wetlands of the Prairie Pothole Region range from freshwater ponds and marshes with ephemeral and temporary water regimes to more permanent, fresh and saline lakes, as well as riverine wetlands. They range in size from < 1 acres (0.5 ha) to > 12,350 acres (5000 ha), although the vast majority are < 2.5 acres (1 ha) with average water depths of < 3 ft (< 1 m). By some estimates, the number of wetlands throughout the entire Prairie Pothole Region is upwards of 9 million (M. Goldhaber, pers. communication).

Playa wetlands have discrete clay basins, and are typically < 3 ft (i.e., <1 m) in depth, and range in size from < 2.5 acres (1 ha) to > 740 acres (300 ha) in surface area (Smith 2003). The average size (surface area) is 15.5 acres (6.3 ha) and most are less than 30 acres (12 ha) in size. There are an estimated 60,000-80,000 such wetlands in the US Great Plains (encompassing portions of Colorado, Kansas, Nebraska, New Mexico, Oklahoma, and Texas), with approximately one-third of these in Texas alone. Playas are the primary source of aboveground freshwater for wildlife in this region and are the primary source of recharge for the Ogallala Aquifer (Bolen et al. 1989). But playas also represent critical sources of biodiversity, accounting for approximately 350 different plant species (Haukos and Smith 1997) and providing critical migration and wintering habitats for nearly 200 species of birds.

Importantly, hydrologic functions, including timing and duration of water inundation of playas, are the result of interactions between climate, topography, soil and vegetation cover, and land-use patterns (Haukos and Smith 1994, Smith 2003, Tsai et al. 2007, 2010).

In the conterminous US, an estimated 50% of wetlands have been lost in the past 200 years (Dahl 1990), mostly in the Great Plains. Most of the remaining wetlands in the Great Plains are intermittent, so any organisms present must have withstood a selective filter for adaptation to ephemeral habitat resources. These wetlands form a naturally spatially heterogeneous and temporally dynamic system, which is under intense anthropogenic demands from agriculture and expected to be acutely impacted by climate change. Under current climate conditions, Great Plains wetlands go through frequent, naturally occurring but unpredictable, wet-dry fluctuations (Haukos and Smith 1994). The timing and duration of these fluctuations is critical to their ecology and delivery of ecosystem services.

STREAMS AND RIVERS

Great Plains streams and rivers are among the most fragmented freshwater systems in the United States (Sabo et al. 2010). This fragmentation is associated with extirpation and reduced population level of some fish (Perkin and Gido 2011). The combined effects of water diversions, impoundments, and increasing water temperatures are likely to threaten many of the remaining species in the Great Plains region.

Habitat fragmentation and flow regulation, which reduce the amount of water in streams for agricultural and domestic uses and often leading to zero flow in both large and small rivers in this region, have contributed to declines in the abundance and distribution of native stream-dwelling fauna (Fahrig 2003, Helfman 2007). Within the contiguous United States, 85% of rivers are fragmented by impoundments that disrupt organism movement and alter stream flow (Huges et al. 2005). These disturbances are associated with the declining and imperiled status of approximately 40% of North American freshwater and diadromous fish (Jelks et al. 2008). Habitat degradation and population effects associated with fragmentation of river habitats include altered geomorphic processes and flow regimes, alteration of dispersal dynamics and isolation of sub-populations, altered phenology and reproductive cues, and overall reduction in amount of contiguous habitat (Gido et al. 2010). Among the principal factors, flow regime alteration is most commonly implicated in the decline of stream-dwelling fish populations, and a growing body of literature suggests flow regime is a major component required for maintaining integrity within stream fish communities (Baxter 1977, Poff et al. 1997, Marchetti and Moyle 2001, Lytle and Poff 2004, Propst and Gido 2004, Taylor et al. 2008, Gido et al. 2010). For example, the magnitude of floods and high-flow pulses that maintain in-stream habitat are reduced following impoundment (Richter et al. 1996, Perkin and Bonner 2010) and, depending upon reservoir management, downstream reaches of impounded streams may experience reductions in mean annual flow and an increase in the number of days with zero flow (Bonner and Wilde 2000). As water availability fluctuates, due to weather and climate, and human demands increase, water reserved for in-stream habitats and species will be more heavily contested and restricted, making flow regimes a critical concern for conservation of Great Plains fish under future climate scenarios.

These patterns of decline transcend spatial scales (i.e., the entire Great Plains), and include multiple levels of phylogeny (i.e., 4 genera, 16 species, 2 subspecies) (Platania and Altenbach 1998, Durham and Wilde 2009a, Gido et al. 2010). The relative abundance of extirpated populations among eight species of suspected or confirmed pelagic-spawning cyprinids is positively correlated with minimum fragment length, indicating that stream fragmentation has played a role in observed declines in abundance and distribution.

Scarcity of water resources on the western prairies, as well as the Western water law doctrine of "prior appropriation," often pits human needs in conflict with each other and with environmental conservation. Sustaining river flows is a fundamental requirement for the persistence of Great Plains fish and other aquatic species, but this water is also coveted for agricultural and domestic uses. As the global climate changes, many models have indicated the propensity and duration of drought on the Great Plains could increase. Increased droughts will increase the probability of conflict between anthropogenic demands and aquatic species requirements, just as the need to maintain healthy habitats increases to support adaptation to uncontrollable changes, such as climate. Connections between land-use practices, wetlands, surface water and groundwater extend the importance and relevance of water availability and water use beyond aquatic environments. The condition and distribution of upland habitats and native grasslands have implications for biodiversity, wildlife conservation and water quality and quantity across the entire Great Plains region. Collaborative, regional efforts have emerged to develop and support opportunities for cooperation and coordination, supporting financial efficiency and regional planning. Increased public knowledge of environmental issues is critical for the continuing success and expansion of these programs.

Responses of Wildlife

BIRDS

Grassland birds are the most consistently declining of all groups of North American avifauna, with 48% of species being of high conservation concern (North American Bird Conservation Initiative - U.S. Committee 2011). These declines have been attributed, in large part, to land conversion and the intensification of agriculture, making the critical bird habitat in the Great Plains among the most threatened landscapes in North America. The population declines likely will be exacerbated by climate change as vegetation and invertebrate food resources respond to altered precipitation, warmer temperatures, and higher rates of evapotranspiration that are expected across the nation's grasslands (North American Bird Conservation Initiative - U.S. Committee 2010, 2011). The different responses among species to environmental change suggest that present-day species assemblages will reconfigure as individual species respond uniquely to the same perturbations.

Wetland-dependent birds, such as waterfowl, shorebirds, wading birds, and ripar-ian associates, are another important component of avian biodiversity in the Great Plains.

Projected temperature and evapotranspiration increases will undoubtedly strongly impact wetland ecosystems and dependent species, several of which are considered to have medium or high vulnerability to climate change, including Western Grebes (Aechmophorus occidentalis), Clark's Grebes (A. clarkii), and Northern Pintails (Anas acuta) (North American Bird Conservation Initiative - U.S. Committee 2010). The shallow depressional wetlands in the playa and prairie pothole regions of the south-central and northern Great Plains, respectively, are acutely threatened by climate change impacts on water levels and sedimentation from upland erosion (North American Bird Conservation Initiative - U.S. Committee 2010, Johnson et al. 2011, Burris and Skagen 2012).

AMPHIBIANS AND REPTILES

Human activities have affected several species of amphibians and reptiles during the past century. At the eastern and northern margins of the Great Plains, wetland drainage and commercial harvesting have severely reduced populations of the northern leopard frog, *Lithobates pipiens* (Koons 1992, M.J. et al. 1994). Prairie streams, important habitats for leopard frogs (*L. pipiens* and *L. blairi*) in drier portions of the Great Plains (Lynch 1978), have been greatly altered by land-use practices (Dodds et al. 2004). In addition, large areas of terrestrial habitats have been degraded or lost, likely influencing the persistence of some native reptile species (Gibbons et al. 2000). Future climate change may affect distribution of amphibian and reptile species indirectly by altering habitat availability, or directly by affecting population demographic characters. There is some evidence for climate-related extinctions of lizards in Mexico (Sinervo et al. 2010), although the impact of climate change on reptiles will likely vary by species. Although climate change effects on amphibians are also diverse (Corn 2005), populations in the Great Plains are less likely than reptiles to benefit from warming temperatures, yet the benefits and costs of climate change to lizards are also poorly understood.

FISH

Stream size is the most important environmental factor determining fish distributions (Schlosser 1982, Fisher and Paukert 2008), however, stream habitat and fish assemblages throughout the Great Plains are not uniform (Matthews 1988) and substrate composition and in-stream cover also play important roles in structuring regional fish assemblages. Large streams and rivers of the region are typically broad, shallow, and often braided with sandy substrates and elevated levels of dissolved solids (Matthews 1988). Riparian cover of narrower streams' canopy is often higher, increasing thermal cover. These physical attributes are important determinants of species distribution across the region. For example, the presence and abundance of the Arkansas darter is associated with narrower streams containing an abundance of in-stream cover (Haslouer et al. 2005), and the plains topminnow is strongly associated with small streams with abundant plant cover (Fisher and Paukert 2008). Furthermore, extensive and sometimes intensive agricultural operations in the watersheds that feed into the Great Plains rivers (Missouri, Platte, Arkansas, Republican/Canadian and Red) provide measurable loading of sediments and contaminants, including nitrogen, phosphorus, and pesticides and herbicides that degrade water quality and habitat conditions (Huntzinger 1995). Extreme events are forecast to increase in magnitude and frequency in several climate models, and these events typically trigger increased rates of overland flow as precipitation rate exceeds infiltration rate. Case studies indicate a two- to three-fold (2-3x) increase in contaminants due to runoff after storm events (Ellis et al. 1984, Staver et al. 1996).

Beyond the general class and characteristics of a stream reach, reproductive success of pelagic-spawning cyprinids is dependent on stream discharge to initiate spawning (Durham and Wilde 2006, 2009b) and to retain eggs in suspension long enough for hatching (Bottrell et al. 1964) and larval fitness and survival, which is a critical population bottleneck (Wilde and Durham 2008, Durham and Wilde 2009b). Thus, the timing and volume of spring runoff and mid-season flows, which are the product of weather and land use, have important implications for the survival of these species within a watershed. Extirpation of pelagic-spawning cyprinids has been greatest in the central and southern Great Plains regions, correlated with notable reductions in discharge since at least the 1970s (Cross et al. 1985, Pigg 1987, 1991, Gido et al. 2010). Further, these same regions include stream fragments created by desiccation, where water does not flow for a majority of the year.

These impacts are chronic but not irreversible. However, climate-induced water limitations and drought will magnify the effects of increasing water demand, making species and habitat conservation dependent upon securing in-stream flows during low-water years. Even when sufficiently long reaches are provided, i.e. > 85 miles (140 km), declining populations of the majority of pelagic-spawning cyprinids were extirpated (73%) of occurrences when stream discharges were reduced by at least half (Gido et al. 2010). Consequently, the possibility exists that discharge reductions, related to anthropogenic withdrawal and climate change, will contribute to declines and extirpations among Great Plains pelagic-spawning cyprinids (see Taylor 2010) and other fluvial organisms, notably fish (Poff and Zimmerman 2010). In the US, 70 species of mussels and 32 species of snails are federally listed as endangered or threatened (U.S. Fish and Wildlife Service 2005).

INVERTEBRATES

Freshwater ecosystems are among the most imperiled ecosystems on Earth: globally, freshwater biodiversity is declining faster than in any terrestrial ecosystem (Revenga et al. 2005). Owing to their short generation times, macroinvertebrates, such as insects, should be particularly sensitive to changes being elicited by our changing climate. Of the invertebrates that have been used as indicators of climate change effects, dragonflies and damselflies (Insecta: Odonata, dragonflies and damselflies) have figured prominently (Samways 2008). These dragonflies and damesflies serve as umbrella species for overall wetland conservation (Oertli 2008), and are one of the insect groups being used to test climate projections (Oertli 2008). Climate change may already be eliciting effects in these insects' distributions and life history characteristics (Flenner and Sahlén 2008, Hassall and Thompson 2008). For example, range shifts attributed to climate change have been documented for dragaonflies in the U.K., with distributions moving higher in latitude and altitude in recent years for several species (Hickling et al. 2005). Phenological shifts have also been noted in the timing of emergence (Hassal et al. 2007). A recently developed North American data warehouse for these insects' distributional data (over 300,000 vetted records from professional and citizen scientists; www.odonatacentral.org), however, will allow us to use a time-series of data that are necessary to distinguish natural variability from trends generated by climate change.

Although overall productivity can be quite high, invertebrate diversity in prairie wetlands is comparatively low because the abiotic conditions are highly variable and often harsh (e.g. Euliss, N. H. et al. 1999, Tangen et al. 2003). Invertebrate community composition is influenced to a large degree by hydrology (e.g., hydroperiod), salinity, and vegetative structure. A majority of the invertebrate taxa are quite resilient to these harsh and variable conditions. For example, ephemeral wetlands that hold water for only a few weeks per year are inhabited by specialized invertebrates capable of completing their life cycles very rapidly, and highly saline wetlands are dominated by taxa with mechanisms for maintaining their osmotic balance. Under the more extreme conditions, however, diversity is often low (Swanson et al. 1988, Euliss, N. H. et al. 1999, Gleason et al. 2009). Invertebrate taxa that inhabit prairie wetlands are generally hardy and thus may be somewhat resilient to direct impacts of climate change.

Because hydroperiod is a well-documented driver of the abundance and distribution of numerous aquatic species (Williams 1997, 2006) that are predicted to be radically altered by climate change, impacts should be seen in the population dynamics and community structure of animals occupying lentic habitats. With predicted changes in precipitation timing and amounts, snowmelt timing, and temperature, changes in wetland water budgets will result in altered hydroperiods and salinity levels and, in turn, may affect invertebrate community structure and productivity. For example, in the northern Great Plains, increased precipitation could extend hydroperiods and indirectly affect invertebrate productivity by moderating the nutrient cycling normally promoted during drying periods. Extended hydroperiods, elevated water depths, and increased wetland connectivity also could result in conditions that are more favorable for colonization by fish, which have been shown to impact ecosystem structure and aquatic invertebrate communities (Zimmer et al. 2000, Tangen et al. 2003, Hanson et al. 2005). Lastly, fluctuations in snowpack and temperature may affect the timing of the preliminary spring hatch of invertebrates associated with the smaller, seasonal wetlands.

Management Opportunities and Challenges

The dynamic nature of climate has long been an issue of duality, where land managers simultaneously recognize the inconsistencies in weather (rainfall, drought, etc.), but neglect moderate to long-term considerations of weather patterns for guiding our understanding of the systems and planning for future management. This is true, in part, because the planning horizon for most units is ten to twenty years. But long-term perspectives, along with forecasts and observations, indicate that rapid changes and extreme variations in weather are possible, even within these planning horizons and certainly into longer term considerations.

From a management perspective, whether the focus is commodities, or conservation of species, the dynamics of climate represent yet another uncontrollable variable affecting health and productivity of systems. This puts climate change in a dubious category, along with land use, resource extraction, pollution and economic production, as factors and forces that contest or challenge sustainability of operations and conservation of species and wild habitats. The primary underlying drivers that challenge conservation of ecosystems and biodiversity in the face of climate change include alteration of freshwater systems, land use intensification (especially conversion of terrestrial and wetland systems to agriculture and domestic purposes), habitat fragmentation (division and isolation of remnant natural systems), and modification of natural processes such as fire and herbivory. The relatively low proportion of land protected for conservation clearly indicates that conservation will be effective only through broad-scale partnerships that will likely include public, private, and non-governmental organization parties.

Despite the relatively small area with protected status, national parks in some basins of the Great Plains host the majority of remaining native fish species (Lawrence et al. 2011). On one hand, these relict populations offer hope for conservation and expansion of native species to all or part of their former ranges. On the other hand, these fragmented relicts might represent the survivors of an anthropogenically induced bottleneck. If the latter case is reality, observing the subsequent extinction and/or fitness of species as these new populations are tested by climate variability will provide an informative, but potentially gruesome, evolutionary experiment. The interaction of natural and anthropocentric management of grasslands, and former grasslands (i.e. agriculture and urban) across the Great Plains promises to be challenging and contentious. Agriculture and other intensive land uses destabilize the soil profile and enable transport (loss) of critical nutrient and water retention capacities (Samson et al. 2004).

Therefore, opportunities for conservation of native grasslands, including species and processes, lie primarily and most immediately on a fragmented network of untilled prairie. Most of these lands continue to receive intensive use, especially from domestic grazing. These systems developed with significant grazing pressure, but the historic herds of the Great Plains adapted to climate, disturbance and associated habitat variability by migrating (Samson et al. 2004). Modern land-use patterns and structures, however, preclude landscape-scale migrations. It will be difficult to restore these large-scale processes across the region, but restoration of processes, conservation of remnant species and habitats, and consolidation/connection of fragmented areas at landscape and local scales will be necessary to provide conservation of species and ecosystem services across the region. New adaptations and flexibility is needed at the interface between native habitats and ecosystems and agriculture.

Recent history is characterized by sod-busting, wetland draining, and open-range fencing, but relatively little emphasis and effort have been placed on restoration of abandoned prairies. The realities of climate change and scarce groundwater supplies promise to force change on institutional relationships and infrastructures that attempt to restrict and restrain natural variability. Managers must bring a renewed emphasis on soil and wetland restoration, not simply dumping refined sewage on degraded soils or manufacturing retention ponds, but restoring species and processes that provide critical ecosystem services, including soil stability and health, water conservation, aquifer recharge, and forage for wildlife and domestic herbivores. In turn, these species and processes can support a sustainable socio-economic system where local products, tourism and culturally significant species accompany large-scale agriculture, industry, and international trade as fundamental components of society. Although industry and investment bankers prefer structure and stability (due to perception of strength and insurance), social-ecological systems in this region, and likely elsewhere, must embrace dynamics and adapt.

Successful adaptation of human systems and conservation of natural systems, with any semblance of healthy function will require (1) vision and regional scale planning and implementation, (2) renewed emphasis on restoration of ecological systems and processes, (3) recognition of the value, importance and "reality" of natural dynamics and diversity, and (4) considerable "luck" because changes, such as extinctions, can occur rapidly when populations are small and mobility is restricted. While ecological understanding has expanded tremendously in the past 100 years, we still know very little about many of these species and systems.

Chapter 6

Energy Considerations

Climate and Energy Context for the Great Plains Region

There is strong seasonally dependent variability in both energy and water demand in this region. Water use peaks during the summer irrigation season, with the timing of greatest water consumption dependent on cropping patterns and constraints on water availability (Schneekloth & Andales, 2009). Total energy use for irrigation water delivery varies considerably across the region because of differences in total cropped acreage, dependence on groundwater or surface water, depth to groundwater, type of crop grown and weather-dependent crop evapotranspiration (e.g. USDA National Agricultural Statistics Service 2010). Overall, electric power use generally follows a U-shaped seasonal pattern with higher consumption in both winter and summer than in spring or fall (Colby & Tanimoto, 2011; Fan, Methaprayoon, & Lee, 2007; Fisher & Ackerman, 2011). A statistical analysis of the role of weather variables in driving seasonal differences in electricity demand in Arizona, found that: "...the relationship between load and temperature follows a quadratic pattern... temperature levels that are far from a certain neutral point lead to more consumption of electricity for cooling or heating. ... the insensitive level found in exploratory analysis was around 59 °F (15 °C)" (Colby & Tanimoto, 2011).

A similar U-shaped relationship between electricity use and temperature was found in the Midwest with the exact shape of the relationship varying somewhat between different areas because of differences in the sensitivity of demand to weather (Fisher & Ackerman, 2011). Similarly, differences in the exact shape of the temperature, electricity-demand relationship were found across states (Fisher & Ackerman, 2011) and across different climatic zones within a single state (Aroonruengsawat & Auffhammer, 2009). In all of those studies, winter heating and summer cooling demand were found to be important drivers of seasonal electric use variability. These findings suggest that summer cooling demand is likely to become an increasingly important driver of electric power use in the Great Plains as the region's climate warms, as evidenced by record peak electric power use in Texas during the record heat wave in the summer of 2011 (Electric Reliability Council of Texas, 2012).

Climate change that will result in increased summer extreme heat days will require more of both energy and water in the Great Plains region. This will be discussed in more detail below.

Overview of Energy-Water-Land Nexus

Most forms of energy production require significant amounts of water for mining, fuel processing, and electric power generation (Averyt et al., 2011; Cooley, Fulton, & Gleick, 2011; U.S. Department of Energy, 2006). In addition, moving and treating water consumes major amounts of energy, especially in areas where it has to be moved great distances from the source to the users. As a result of multiple, interacting stressors at the water-energy nexus, the Great Plains region is experiencing increasing problems with both water quality and quantity for maintaining critical ecosystem services such as drinking water, irrigation for crops, hydropower, healthy fish populations, aesthetic and spiritual values, and many others. These stresses are especially common in the semi-arid western areas of the region, which face even drier conditions with climate change, and along many major rivers systems, which are already over-allocated to agricultural, municipal, industrial, recreational, and environmental uses (Barnett & Pierce, 2009).

Decisions made today about water and energy use and climate adaptation and mitigation will have impacts for decades to come. The myriad uncertainties posed by alternative socio-economic pathways and different plausible climate change scenarios mean that decisions taken today need to take into account the risks of climate change and these multiple stressors in the future.

Land use and land-use changes are closely linked to the availability and use of water resources and energy. Energy demand and the resource and economic opportunities for developing renewable and non-renewable energy resources, such as gas and oil, coal, biofuel, hydropower, solar, and wind power, are high. Energy production, including alternative-energy options, have a wide range of effects on land cover and productivity, and also impacts other factors that affect carbon, water, and energy fluxes and, in turn, climate (Dale, Efroymson, & Kline, 2011). Relative energy impacts on land use are influenced by characteristics such as the extent, duration, intensity, and reversibility of change. Energy infrastructure for storage, transportation, and processing will likely alter the landscape for long periods of time. Likewise, conversion of native prairie grasslands to croplands is almost irreversible since these lands' ecological integrity has evolved over thousands of years.

At the heart of the issues bridging energy, water, and land is the nexus between climate mitigation and adaptation. Mitigating emissions of GHGs has implications for both water and land resources. Practices that include evolving fuel portfolios, carbon capture and storage technologies, and land sequestration of carbon have the potential to compromise our ability to adapt to climate-driven impacts to water and land resources. Similarly, in an effort to adapt to changing water and land regimes, moving water and altering land can be energy intensive—creating a feedback that may compromise efforts to minimize greenhouse gas emissions.

The energy-water-land nexus is a multi-stressor problem with drivers that extend beyond the climate. Population growth and concomitant demands for energy, municipal water supplies, and land are also concerns. Texas and Wyoming were among the states that saw the largest percentage of population growth since 2000 according to the latest US Census. Austin, Texas was among the top ten fastest growing metropolitan centers in the nation (37.3%), and Lincoln County, South Dakota was one the most rapidly growing counties (85.8% (U.S. Census Bureau, 2011)) (See Table 6.1). Translating population or population growth into water demands and land use, however, is not straightforward. Water and energy demand are not directly related to population or population increases, largely because of conservation efforts. Land use is similarly difficult to correlate. For example, in 2008, Texas, Nebraska, and Kansas came in second, third and

Table 6.1 Population Growth 2000–10 in Great Plains Region

Kansas	6.1%	Oklahoma	8.7%
Montana	9.7%	South Dakota	7.9%
Nebraska	6.7%	Texas	20.6%
North Dakota	4.7%	Wyoming	14.1%

Source: (U.S. Census Bureau, 2011)

fourth, respectively, behind California, for total on-farm energy use for irrigation pumping. These three states account for 40% of total energy use (by expenditure) of the nation's use of power (of all types) for irrigation -- just over $ 1 billion out of a national total of ~ $2.68 billion (USDA National Agricultural Statistics Service, 2010). Navigating the nexus is expected to become more difficult as the regional climate continues to change.

Water for Energy

Water is required for the development of most energy resources: from extraction, to building infrastructure, to generation of electricity. The thermoelectric cooling process (where water is used to spin a turbine to generate electricity, and is then cooled) accounts for a greater proportion of national freshwater withdrawals than agriculture (U.S. Geological Survey, 2009). But different combinations of fuels and cooling processes use different quantities of water (Figure 6.1 below) (MacKnick, Newmark, Heath, & Hallet, 2011). For each kWh of electricity generated, nuclear technologies withdraw and consume the most water. Water use associated with concentrated solar plants is also relatively high, on par with coal-fired power plants. An important point here is that low carbon does not always equal low water use.

There are approximately 1,750 power plants in the Great Plains. In 2008, these plants generated (2,300 10^3 GBTUs/674,000 10^3 GWh) of electricity primarily using coal (50% of total generation) and natural gas (34%). The dominant cooling technology was once-through cooling—meaning that heat is dissipated through evaporation and hot water is not discharged back into rivers, streams, and lakes. The combination of power plants in the Great Plains yields a water intensity of 6.4 gallons (24 liters) of water withdrawn per kWh generated, and 0.4 gallons (1.5 liters) consumed per kWh. However, there are variations from state to state (Figure 6.2, (Averyt et al., 2011)). In contrast with the national portfolio, agriculture is the largest water user in the Great Plains region. Although much agricultural water is drawn from groundwater resources, 96% of water for thermoelectric cooling comes from known surface water sources, and less than 1% comes from groundwater. Aside from water for power plants, water use for energy development

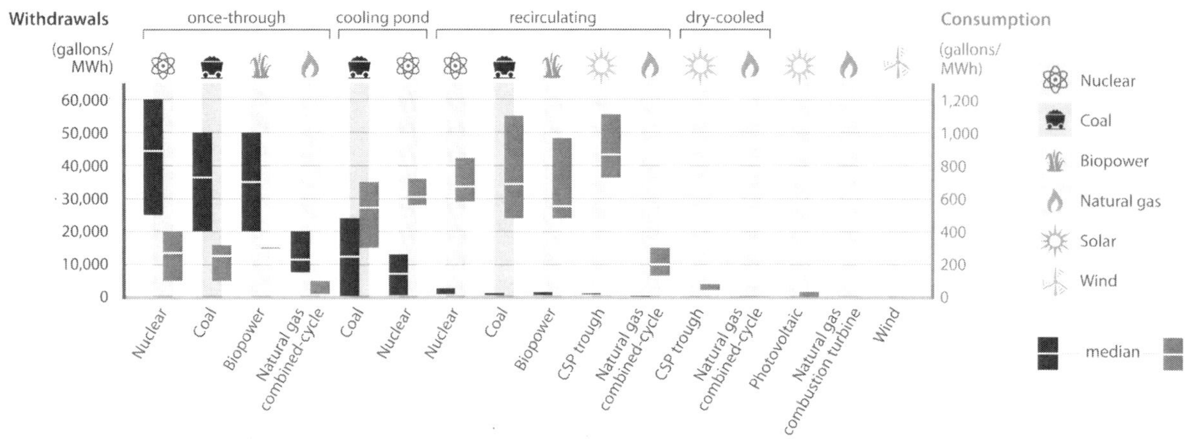

Figure 6.1. Caption from report: Water withdrawals per megawatt-hour (MWh) can range from almost zero for a solar photovoltaic, wind, or dry-cooled natural gas plant, to hundreds of gallons for an efficient plant using recirculating cooling, to tens of thousands of gallons for a nuclear or coal plant using once-through cooling. Water consumption per MWh can similarly range from almost zero for solar, wind, or gas plants using dry cooling to around 1,000 gallons (3,785 liters) for coal, oil, or concentrating solar power (CS P) with recirculating cooling. How much water a specific plant uses reflects its efficiency and age, and how much the plant is used, along with local humidity, air temperature, and water temperature (Averyt et al. 2011, MacKnick et al. 2011)

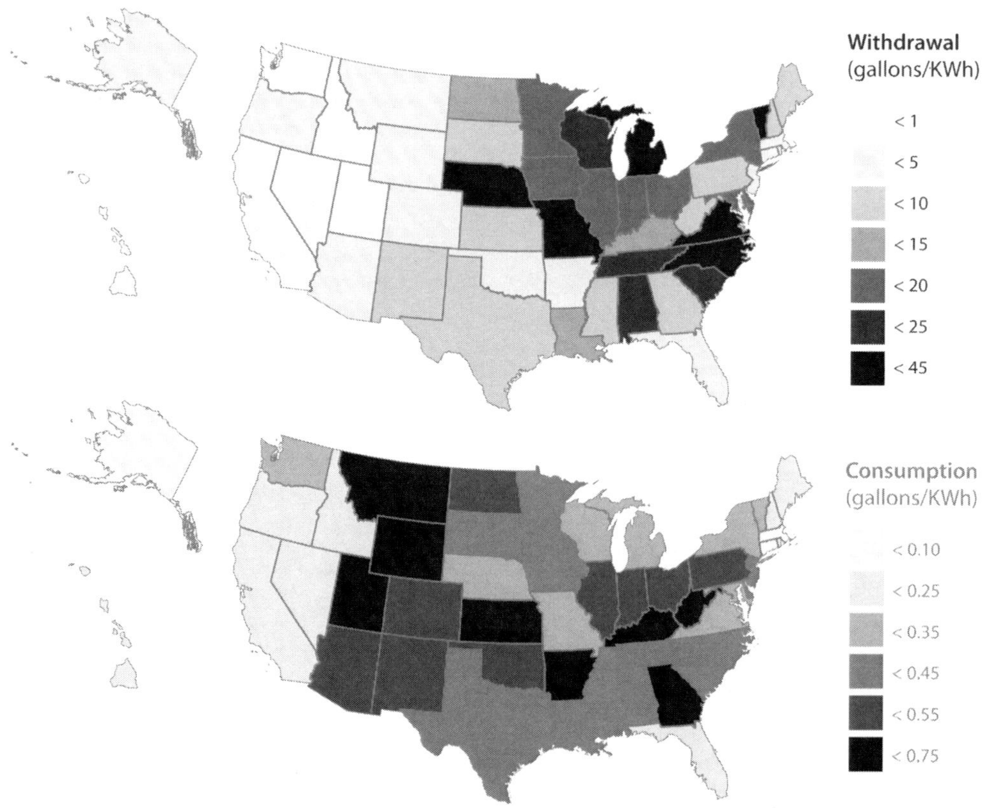

Figure 6.2. Freshwater Use for Electricity Generation by State (Averyt et al. 2011)

and the implications for water quality are issues in the Great Plains. Water requirements for most extraction practices are ill constrained and highly variable.

Energy for water

Energy is required to pump, treat, distribute, and use potable water, and to treat and discharge wastewater. The *energy intensity* of water, or the energy used to provide a unit of water (e.g., a gallon, acre-foot), depends on the source and quality of the raw water, and the type of use. For example, pumping raw water over long distances or over mountain ranges can use a large amount of electricity; California's State Water Project and Arizona's Central Arizona Project are well-known examples. Many cities in the West rely on high-quality water that flows to city treatment plants by gravity, requiring very little energy to pump, treat, and distribute the water to customers. Increasing urban water supplies will, in many cases, require cities to pump water over greater distances or from deeper aquifers.

The energy intensity of water will vary depending on the source (i.e. surface or groundwater) and the quality of the water. Cities that rely on surface water fed from snowmelt in the Rocky Mountains (e.g. Denver, Colorado) generally require only moderate amounts of energy to treat and distribute water. For example, the energy intensity of treating and distributing water in Denver, Colorado, in 2007 was 0.8 million BTUs/AF (188 Wh/m^3)[2] Colorado Springs, Colorado has also relied primarily on gravity-fed water supplies from the Rocky Mountains. To expand its supplies, Colorado Springs recently began construction on the Southern Delivery System, a project that will pump water from Pueblo Reservoir to Colorado Springs, requiring an estimated 16 million BTUs/AF (3750 Wh/m^3) (not including treatment or distribution) (Figure 6.3). In many parts of the West, water demands already exceed supplies, creating a need to import water between watersheds and across state lines, and tap additional groundwater resources (U.S. Bureau of Reclamation, 2008a).

Water providers are developing even more water supplies that require pumping from greater depths (groundwater) or conveyance over longer distances. In the future, water providers may also need to increasingly rely on lower quality supplies that require more extensive treatment, such as tapping more saline supplies that require reverse osmosis. The energy intensity of reverse osmosis depends on the salinity of the water treated. For example, in its 2007 demonstration run, the Yuma Desalting Plant used approximately 5 million BTUs/AF (1.2 kWh/m^3) to treat brackish water (salinity of 2,539 mg/L, reduced to 252 mg/L) (U.S. Bureau of Reclamation, 2008b).

In addition to changing water availability, climate change may affect the timing and magnitude of runoff. For many water utilities, existing storage facilities may adequately accommodate variable runoff regimes. Some utilities, however, may require additional storage. If "new" storage includes aquifer recharge (and subsequent recovery), this may add pumping processes resulting in additional energy demands. Finally, wastewater treatment plants often discharge treated wastewater into streams, depending on adequate stream flows to ensure that discharges do not exceed stream temperature or water quality standards. Reduced stream flows or elevated stream temperatures may

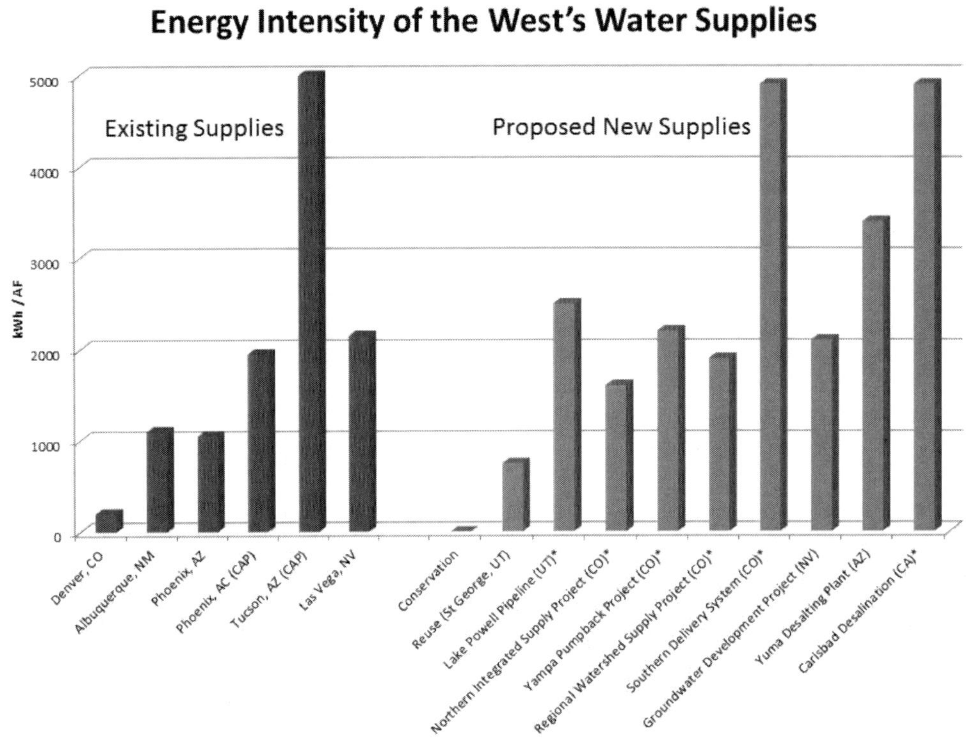

Figure 6.3. The energy intensity of many proposed projects exceeds the energy intensity of existing supplies. Notes: *Figures include an estimated 150 kWh/AF for treatment and/or distribution. +The Yuma Desalting Plant includes the energy used on site and the energy used to pump water to participating utilities in Arizona, Nevada, and California, as its operation is designed to increase water supplied to cities in those states. Colorado Springs' Southern Delivery System and the Carlsbad Desalination Plant are now under construction. The upper map only includes the Colorado River system. These different projects require varied quantities of energy (adapted from Spears et al. (2011)

drive wastewater treatment plants to increase treatment standards, elevating the energy intensity of treatment.

Managing the impacts of diminished and changing water supplies can be informed by current adaptation strategies. New water supply projects such as Colorado's Southern Delivery System may increase and diversify a water utility's water supply portfolio, but could also increase total energy demands. Alternative options include water conservation, increasing use of recycled water, and developing flexible leasing arrangements between cities and farmers. Each of these options has different benefits. Water conservation can both reduce total water demands and save energy, particularly if conservation efforts focus on reducing the use of hot water and/or energy-intensive water conveyance or pumping systems. Recycled wastewater is typically drought resistant, but, depending on the level of treatment required to provide recycled water, it may have additional energy demands. Ultraviolet disinfection, for example, is energy intensive. However, the energy used to treat and distribute recycled water may be less than the energy required for new water supply projects. Under traditional agricultural-urban leasing agreements, cities pay farmers to temporarily fallow a portion of irrigated agricultural land and transfer water to cities. These agreements may enable cities to mitigate the impacts of more extreme droughts without increasing the need for energy intensive new infrastructure projects.

The energy impacts of adapting to changing water supplies are an important consideration. Some of the strategies described above may help cities both adapt to and mitigate climate change, while others help cities adapt, but increase GHG emissions. The energy requirements necessary for adapting to climate-driven changes in water supply is an example of how decision making about climate adaptation can come into conflict with efforts to mitigate greenhouse gas emissions.

Energy Options and Tradeoffs - Different Effects on Water and Land

The Great Plains region has an abundance of coal with high potential for development (U.S. Geological Survey, 2003). The most productive coal mines in the country are in the Powder River Basin of Wyoming (Averyt, 2011). However, coal-fired electric power plants are not only a major source of the GHG carbon dioxide and other air pollutants, but they are also heavy users of water (Averyt et al., 2011). Climate change will likely decrease water availability in already stressed areas and create increased competition among users. Rural communities often face sparse economic opportunities and many communities are highly dependent on jobs and tax revenues from fossil fuels – predominantly coal in the Powder River Basin area. This creates major challenges and tradeoffs for their efforts to develop their economies and chart sustainable livelihoods, especially as the nation and global community transition to a cleaner energy future.

Choices about how we produce electricity in the coming decades could have a big impact on water consumption. For example, if the nation were to get 20 percent of its electricity from wind by 2030, water consumption could be reduced by about 10 percent, compared to 2005 consumption. On the other hand, if carbon capture and sequestration (CCS) technologies are widely adopted, water consumption could be increased further by 7.5 to 19 percent since CCS uses cooling water for the capture and compression processes and to generate the extra electricity needed to perform CCS. Developing concentrated solar power plants also presents tradeoffs between water consumption and power generation efficiency, especially if dry-cooling approaches are used in hot climates. Because some of the electricity generated must be used to operate fans, electricity from a dry-cooled plant can cost about 10 percent more than that from a wet-cooled plant. These effects are especially acute when ambient temperatures exceed 100°F (38°C). Hybrid wet-dry cooling approaches are currently being developed as a promising alternative. These systems use dry cooling unless temperatures exceed a certain threshold, at which point they switch over to evaporative cooling. Such systems can use 90 percent less water than plants that rely only on evaporative cooling, and only see a 3 percent drop in energy performance. A potential source for cooling water in the Great Plains region is the usage of treated municipal wastewater.

Vulnerabilities and Mitigation/Adaptation in the Context of Future Energy-Water Demand and Supply

Capital investments for resource infrastructure, such as reservoirs and power plants, represent large-scale and long-term resource commitments, which are difficult to reverse once set in motion (Hegmon et al., n.d.; Scheffer & Westley, 2007). Iterative risk-based management and adaptive governance approaches are necessary for sustaining water and energy resources while maintaining sustainable livelihoods in the face of increasing demands for both. Evaluation of these tradeoffs between agriculture, energy, municipalities and the environment are needed to better assess the appropriate strategies to be considered.

Delivering water and wastewater services is an energy-intensive effort, as the water is treated, pumped to homes and businesses, then pumped to wastewater facilities to be treated again. EPA estimates 3-4 percent of national electricity consumption -- equivalent to approximately 53 billion BTUs per second (56 billion kilowatts) -- is used in providing drinking water and wastewater services each year. Pursuing energy efficiency through these systems can significantly reduce operating costs, while mitigating the effects of climate change. Numerous resources exist to help water utilities pursue efficiency measures, including EPA's Energy Management Guidebook for Wastewater and Water Utilities (U.S. Environmental Protection Agency, 2008), which is part of the agency's Sustainable Infrastructure effort (U.S. Environmental Protection Agency, 2012a, 2012c). Utilities in the Great Plains have been working with EPA to develop energy management programs based on the guidebook, as well as case studies to demonstrate the benefits that they are seeing. One example in the Great Plains is the Missouri Water Utilities Partnership - Energy Management Initiative for Public Wastewater and Drinking Water (U.S. Environmental Protection Agency, 2012b). The eight participating cities are in various stages of implementing projects that are collectively projected to reduce energy consumption by more than 8 million kWh (7.6 million BTUs per sec), while cutting greenhouse gas emissions by 16 million lb (7.3 million kg).

Energy Transmission

The Great Plains sits at the physical intersection of all three national grid systems, on the seam between the Eastern and Western Interconnects, which divides the Dakotas, Nebraska, Kansas and Oklahoma from Montana, Wyoming and Colorado, and astride the Electric Reliability Council of Texas system in Texas to the south. The Missouri River Basin straddles this "electrical continental divide", with its headwaters in the Rockies of Montana and Wyoming and the bulk of its flow and hydropower generation into the Eastern Interconnect.

These three systems are fully independent, with the east and west flow of power interchanges across the seam through the direct current exchanges in Montana, South Dakota, Nebraska and New Mexico. This situation in the Great Plains makes planning and operation of the electric system across the region more complex than if it was a single system (Kaplan, 2009).

The physical infrastructure of the electrical system in the Great Plains is composed of a variety of generation facilities, including hydropower, coal, gas, nuclear, and

BOX 6.1

Case Study: Texas Drought and Energy-Water Impacts

In 2011, the Southern Plains drought was characterized as a "flash drought" because the onset was so rapid, coming in weeks as opposed to months or seasons. While some portions of New Mexico, Oklahoma, Nebraska, and Louisiana experienced extreme to exceptional drought in 2011, Texas was at the epicenter of the event with the entire state experiencing some level of drought. At the height of the drought in October, over 80% of the state experienced a D5 "exceptional drought" stage (NIDIS, 2011). Many weather stations in Texas showed a mere 25% of the normal 12-month precipitation (Nielsen-Gammon 2011). Accompanying the drought was one of the worst heat waves on record, which resulted in increased evaporation that further depleted already low streamflow and reservoir levels (Nielsen-Gammon 2011). During summer 2011, Texas experienced both the hottest and driest conditions on record: temperatures were observed to be 2.5 °F (1.4 °C) hotter than the previous record set in summer 1980 and rainfall was 2.5 inches (6.4 cm) lower than previous low rain amounts recorded in 1956. Other drought measures attest to the severity of the drought: "Texas' average Palmer Drought Severity Index (PDSI) from June through August, 2011 was -5.37 – the lowest, indicating the most severe drought conditions, since the start of the instrumental record in 1895" (Dawson, 2012). In the long-term paleo-record using tree-ring data, the 2011 drought was matched in severity only in 1789 (NOAA 2011b). The severity of the drought appears to be the product of a La Niña event, exacerbated by climate change (Nielsen-Gammon, 2011).

The 2011 drought threatened thermoelectric generation through limited availability of water while the heat wave induced increased demand for peak electricity. "More than 11,000 megawatts of Texas power generation — about 16 percent of the total power resources of the Electric Reliability Council of Texas — rely on cooling water from sources at historically low levels. If Texas does not receive "significant" rainfall by May 2012, more than 2.8 million BTUs per second (3,000 MW) of this capacity could be unavailable due to a lack of water for cooling" (ERCOT - Electric Reliability Council of Texas, 2011). This potential impact is further intensified when considering that increased cooling demands caused by the heat wave drove peak electricity demands to all-time highs, exceeding the prior record on eight of the first twelve days of August 2011. The peak demand rose to 64.7 million BTUs per second (68,294 MW) closely approaching the state's capacity of 68.2 million BTUs per second (72,000 MW). While the Texas' growing utilization of wind power, currently 12.5% of the state's energy production, reduces challenges posed by limited water supplies, it places the state at greater risk of not meeting peaking demands due to the inherent variability of production.

Beyond thermoelectric generation, limited water can also constrain gas shale production. In 2010, the Texas Water Development Board estimated that 13.5 billion gallons (15.1 billion liters) of water were used in the drilling and stimulation of gas shale wells in Texas. In August 2011, the town of Grand Prairie, in the northern part of the state, became the first in Texas to enact a ban on the use of water for hydraulic fracturing, or fracking (Malewitz, 2011). The Texas Water Development Board acknowledges concerns about the use of water for hydraulic fracturing in the energy industry, and says it will monitor this closely in its next regional water planning cycle (Texas Water Development Board 2011).

renewables (primarily wind). The system includes both the high voltage transmission and the stepped-down lower voltage distribution systems owned and operated by the Western Area Power Administration, rural electric cooperatives, public power districts and municipal utilities. In the Upper Plains region, public power utilities own and operate almost half of the high voltage grid, as compared to the rest of the country where up to 80% may be operated by investor-owned utilities (Kaplan, 2009).

The federal transmission grid was originally built by the US Bureau of Reclamation beginning in the middle of the last century to collect and transmit electrical energy from Reclamation and US Army Corps of Engineers hydroelectric dams in the Missouri River watershed to meet energy demands throughout the upper Midwest and West. Today, the system is jointly operated by Western Area Power Administration, in conjunction with regional generation and transmission organizations as an integrated system, through a complex set of federal authorities and federal and pubic agreements that have been developed over the past 50 years.

Western Area Power Administration is one of four federal power marketing administrations directed by law to market and transmit federal power allocations at cost-based rates to preference customers, including federal and state agencies, rural electric cooperatives, public power districts, Native American tribes, and municipal utilities. This hydropower is delivered through nearly 100 substations, across nearly 7,800 miles (12,550 km) of federal transmission lines ranging from 69 KV to 500 KV in the Upper Great Plains region (Western Area Power Administration, 2012). These lines are connected with other regional transmission systems and groups. The physical transmission infrastructure, and especially the distribution system consisting of tens of thousands of miles of wire on towers and poles, is significantly vulnerable to weather extremes and climate change. The higher voltage is susceptible to short-circuiting during summers due to stretching of transmission wires during periods of overheating caused by overloading and record high temperatures, as well as during winter ice storms.

In addition, winter weather conditions can combine to wreak havoc on the electric cooperatives' power system, where ice clings and builds on the power lines, causing them to sag under the tremendous weight. Blustery winds ripple the already heavy lines, making them "gallop" and eventually cross. Transmission lines on the prairies, where there are little to no physical features to block the wind power lines and poles, are vulnerable to extreme winds or winter storm conditions.

In November 2005, over 1,200 high-voltage transmission poles were destroyed by ice and wind in East River, South Dakota, with 725 miles (1165 km) of transmission lines put out of service and 35 substations serving local distribution cooperative systems taken offline, at a repair cost of $6 million for the transmission system (East River Electric, 2006). The lower-voltage distribution systems are even more susceptible to catastrophic ice storms, such as events in the early winter of 2005 and in the late winter and the spring of 2010 and 2011 (Basin Electric Power Cooperative, 2011). In 2005, many local electric distribution systems were hard-hit by the combination of ice, snow and wind, and an estimated 10,000 distribution poles went down, leaving more than 20,000 electric cooperative members in eastern South Dakota without power as frigid Arctic air arrived. The most widespread and devastating ice storm in the state's history caused an estimated $20 million in damages to the rural electric cooperative systems (East River Electric 2006) In 2010, icing conditions destroyed nearly 20 percent of one electric cooperative's system in North Dakota, requiring the rebuilding of 500 miles (805 km) of line in often very remote areas with rugged terrain over a three week period (North Dakota Association of Rural Electric Cooperatives, 2012).

© Sofi Jaramillo

Chapter 7

Agriculture and Land Management

While Great Plains agriculture is highly productive, rising input costs associated with high energy costs, changing demographics, and extreme climate events decrease its resilience. Until the recent increases in commodity prices, associated with rising demands for exports and bioenergy, farm-gate prices were often below the full cost of production. The long history of subsidies to US farmers has become increasingly controversial, and many people outside the agricultural sector are advocating for a changing U.S. policy, and shifting federal expenditures toward programs supporting payments for environmental services, new farmers, and alternative production practices, such as organic and healthy food programs. Many macroeconomic factors affect the stability and resilience of US agriculture in a global environment, including variability in currency exchange rates, changes in international trade, foreign and domestic income, rural employment, interest rates, and energy costs (Pender, Marré, & Reeder, 2012). Franzluebbers et al. (2011) discussed many of the forces acting on US rain-fed agriculture associated with the Great Plains, including challenges to maintaining soil water, soil protection against erosion, and improving crop-livestock systems to reduce external inputs.

Changing Environmental Factors

Climate variation and extremes have always been a defining characteristic of the Great Plains, and no sector is more vulnerable to climate than agriculture. One aspect that is often under-appreciated is the extent to which multi-year patterns characterize the climate record. For agriculture and many other sectors, multi-year droughts present a more difficult challenge than shorter term droughts, as soil and water reserves as well as financial resources are depleted. Similarly, multi-year wet periods may offer opportunities to intensify production, but also may pose increased challenges due to waterlogged soils, flooding, and diseases.

Similar multi-year patterns are seen with temperature. To some extent, there is correspondence between wet and cool periods and dry and hot periods, but additional factors influence these patterns. Heat waves can cause severe costs and yield reduction for livestock and crop production, over and above the losses often associated with drought.

Extended growing seasons associated with a warming mean climate may present an opportunity to diversify cropping. Crops across the region's diverse landscape will be impacted differently by climate change. Production in some areas will increase due to more rainfall and longer growing seasons, but drought and higher temperatures will cause production to decrease in other areas. Additionally, shifts in precipitation and temperature will influence pests and weeds (Karl, Melillo, & Peterson, 2009).

Critical temperature ranges for life cycle development differ for different crop species, such as wheat, corn, soybean, or cotton. As indicated in Chapter 3, mean air

Table 7.1 Percent grain yield and evapotranspiration responses to increased temperature 2.2 °F (1.2°C), increased CO_2 (380 to 440 ppm), and the net effects of temperature plus increased CO_2. Current mean air temperature during reproductive growth is shown in parentheses for each crop/region as the starting reference.

	Grain Yield			Evapotranspiration	
Crop	Temp. Increase of 2.2 °F (1.2°C)	CO2 (380-440 ppm)	Temp/CO2 Combined Irrigated	Temp. Increase of 2.2 °F (1.2°C)	CO2 (380-440 ppm)
			% change		
Corn[1]	-4.0	+1.0	-3.0	+1.8	--
Soybean[1]	+2.5	+7.4	+9.9	+1.8	-2.1
Wheat	-6.7	+6.8	+0.1	+1.8	-1.4
Sorghum	-9.4	+1.0	-8.4	+1.8	-3.9
Cotton[2]	-5.7	+9.2	+3.5	+1.8	-1.4

[1] Yield and evapotranspiration estimates for the Midwest. 2 Yield and evapotranspiration estimates for the South.
Source: (Karl et al. 2009)

temperatures are predicted to increase across the Great Plains, with variable changes in precipitation. Table 7.1 illustrates the percent grain yield and evapotranspiration response to increased temperature and increased CO_2 (Karl et al., 2009).

Kimball (1983) reported that crop yield is increased by CO_2 fertilization in laboratory and free-air CO_2 enrichment studies, but these yield increases may not be adequate to offset negative effects associated with high temperature and decreased water availability. Some weeds have more positive responses to CO_2 fertilization than most cash crops, particularly cool season, C_3 weeds (Ziska & George, 2004; Ziska, 2003) competing in major C_4 crops (see Box 2.1), such as corn and sorghum. The C_4 weed species show smaller responses to atmospheric CO_2 relative to C_3 crops, but most crops must compete with both C_3 and C_4 weeds and, as weed pressures shift, the industry may not have registered pesticides for new crop-weed combinations. Additionally, the most competitive weeds for a particular crop are those with similar growth habits and photosynthetic pathways, and weed / crop competition studies show weed growth is favored over crops of similar photosynthesis as CO_2 increases (Ziska & Runion, 2006). Ziska et al. (1999)

also suggest that glyphosate herbicide (the most commonly used herbicide in the U.S.) becomes less effective as CO_2 levels increase (Ziska et al., 1999).

Change in CO_2 concentration and climate patterns also impacts beneficial and harmful insects, microbes, and other organisms in agroecosystems. Studies show temperature to be the single most important factor affecting insect ecology, epidemiology, and distribution (Coakley, Scherm, & Chakraborty, 1999). Populations of insect species that are currently marginally over-wintering in high latitude and high altitude regions will increase with warmer winters. Organisms that do not tolerate freezing temperature will move northward. These shifts will lead to an increase in pesticide use, which has ecological effects for other insects and microbes in the area (Karl et al., 2009). An overall increase in humidity and frequency of heavy rainfall events projected for many parts of the United States will tend to favor some leaf and root pathogens (Coakley et al., 1999). However, an increase in short- to medium-term drought will tend to decrease the duration of leaf wetness and reduce some forms of pathogen attack on leaves (Karl et al., 2009). Increased atmospheric concentrations of CO_2 causes higher carbon-to-nitrogen ratios of plant leaves, which can increase insect feeding to meet higher nitrogen requirements (Coviella & Trumble, 1999). However, a diet of high CO_2 plants can slow insect development and lengthen insect life stages where they are more vulnerable to attack by parasitoids (Coviella & Trumble, 1999).

Increased temperatures and decreased rainfall generally affect crops negatively. The degree of harm varies by crop and the point in its life cycle, but temperature increases have the greatest impact when occurring during or just prior to critical pollination phases. A crop's sensitivity and ability to compensate during later, improved conditions, depends on the synchrony of flowering, or anthesis, in each crop (Karl et al., 2009).

The northern and eastern regions of the Great Plains are projected to experience an increase in high-precipitation events (see Chapter 3). An economic consequence of excessive precipitation is waterlogged soils and delayed spring planting, which jeopardizes crops that require long growing seasons. Increased rainfall over concentrated time periods may amplify the likelihood of water shortages at other times due to changes in frequency of rainfalls (Karl et al., 2009). Field flooding associated with intense rainfall events can cause crop losses or yield reduction associated with increased susceptibility to root diseases, anoxia, or soil compaction and crusting; and could also increase leaching of nutrients and agricultural chemicals into groundwater and surface water (Karl et al., 2009). Heavy winds, which often accompany storms with heavy rain, also have potential to uproot crops and reduce yield.

Increased temperatures are anticipated to result in shifting of the cropping patterns across the Great Plains. Beach et al. (2010) evaluated implications of climate change scenarios on the potential range, acreage, and yield of US crops, and found that substantial changes in the distributions and yields can be anticipated for rainfed small grains, hay, corn, cotton, sorghum, and soybean in the Great Plains states. For a range of climate scenarios, agricultural production is projected toward decreased barley production, but increased oats and rye production in the northern Great Plains; a shift of corn toward the south, decreased soybean in northern portions of the region, expansion of cotton to

BOX 7.1
Prairie Heating and CO_2 Enrichment (PHACE) Experiment

As climate change increasingly takes us into a new environmental space, our knowledge and experience from research conducted in present-day and past environments are of limited use for predicting the future. For instance, with ambient CO_2 concentrations now higher than they have been for more than several hundred thousand years, and concentrations predicted to continue increasing, information is needed to understand not only how plants and agro-ecosystems will respond to a warmer atmosphere, but also how rising CO_2 concentrations will affect plants. The PHACE Experiment is one such endeavor to evaluate agro-ecosystem responses to future environments, employing technology to increase ambient CO_2 to 600 ppm and day/night temperatures by 2.7/5.4 °F (1.5/3 °C) to observe how plants and soils of the northern mixed-grass prairie respond to conditions expected in the second half of this century.

Early results from this experiment suggest that the desiccating effects from warming may be offset by considerable improvements in plant water-use efficiency, which occur as CO_2 concentrations increase (Morgan et al. 2011). As a result, average productivity of many native grasslands of the Central and Northern Great Plains may be sustained or even enhanced slightly in the next few decades. However, the possible water-saving benefits are not expected to overcome the severe droughts predicted for regions in more southern latitudes, where both warmer temperatures and declining precipitation are predicted to result in more severe and protracted droughts (Seager and Vecchi, 2010). Thus, the southern Great Plains may experience increased frequency and severity of droughts, curtailing productivity. Further, such CO_2-induced increases in water-use efficiency may eventually be overwhelmed by some of the substantive warming predicted for the end of this century. The PHACE experiment also suggests that rising CO_2 concentrations will not necessarily enhance the ability of such native, semi-arid grasslands to sequester more carbon, in part because the resistant soil C may become susceptible to decomposition under future conditions (Carrillo et al. 2011). These and other results from manipulative type experiments provide important insights of how rangelands will respond to the novel environments we are facing.

the north and east, a decrease of wheat in the southern Great Plains; and an increase of hay across the entire region. Global climate models do not produce reliable information about extreme events, and impacts of extreme events on crop yields are difficult to simulate. However, the losses associated with extreme events are catastrophic, so projections of intensified climate cycles with an increased frequency of extreme events are a concern across the region (See Chapter 3). Rosenzweig et al. (2000) estimated that US crop losses totaled \$56 billion from the 1988 drought and \$23 billion from the 1993 floods along the Mississippi River. The 2011 drought resulted in over \$5.2 billion is agricultural losses in Texas alone (Texas A&M University, 2012), and resulted in massive residential, wildlife, tourism, and agricultural losses due to wildfires.

Opportunities and Challenges and Changing Farm Trends as a Result

Given the great importance of Great Plains food production, research, extension, and policy efforts have been undertaken at federal, state, and private sector levels to improve production, efficiencies, and environmental protection. Many of these modifications have been taken in response to regulatory demands related to conservation practices and good stewardship standards. The improved technologies also contributed to reduced diversity in agricultural systems (Sylvester & Cunfer, 2009). However, a number of practices have also been developed to improve agricultural efficiency practices related to water use, soil tillage, and nitrogen usage. In addition, market changes related to energy and commodity prices have influenced the crop production systems, as evidenced by the expansion of corn production to accommodate the corn-ethanol production industry.

WATER CONSERVATION

A large proportion of Great Plains agriculture is extremely vulnerable to groundwater depletion associated with over-allocation of water from an aquifer with extremely limited recharge. The southern extent of the Ogallala Aquifer has already been depleted by 274 million acre-feet (338 billion m^3) since predevelopment (before 1950) (McGuire, 2011). With all sectors relying on groundwater, and agriculture being the greatest use of water, improved irrigation efficiency or conversion to rainfed production to reduce groundwater extraction has been a long-term focus of research and technology development. There have been dramatic increases in irrigation use efficiency due to better technologies, conversion to more efficient methods (e.g. adoption of drip irrigation), improved irrigation scheduling, and more water efficient crops and varieties. However, land will continue to be taken out of irrigated production because of the failure of wells or sales of water rights to non-agricultural users, and when this occurs, effective conversion of irrigated land to rainfed cropping or perennial vegetation is essential to protect the soils from wind and water erosion and to provide ongoing economic benefits to landowners.

More water-efficient rainfed production in the Great Plains is needed to minimize the economic disruption caused by groundwater depletion and the decrease in irrigated agriculture. No-tillage systems have provided enhanced water conservation and allowed for diversification and intensification of cropping systems. The greatest adoption of no-tillage has been with crop species that have herbicide-tolerant variants. To date, the industry has not been able to ensure good integrated pest management practices with pesticide- and herbicide-tolerant crops with many pest species now exhibiting tolerance to the applied chemicals. This challenge limits the resilience of future crop production systems, as the need for reduced tillage for water conservation will increase.

PRECISION NUTRIENT MANAGEMENT

Recent advances in nutrient and soil water monitoring at the field level has led to improved precision nutrient management practices. Use of the technologies has reduced the amount of nutrients and water applied. The technologies provide adaptive nutrient management strategies that modify nutrient and water application rates, timing, and guidance for recording management practices (USDA National Agricultural Statistics Service, 2010).

Precision nutrient management, which shifts fertility management from whole field to an acre-by-acre basis, increases yields, but it is also requires capital investment greater than traditional practices. In a comparison of site-specific management zones of continuous corn cropping systems in northeastern Colorado with conventional uniform applications, Koch et al. (2004) determined the precision regime is be more economically feasible due to a decrease of total fertilizer inputs but an increase in yield.

Impacts of Climate Change on Great Plains Livestock Industry

There are several potential climate change impacts on livestock production systems. These are primarily determined by impacts on feed production (forage biomass production, forage quality), water availability, animal effects both direct (thermal stress) and indirect (decreased immunity, increased disease and parasites, decreased reproduction or weight gain), and other factors. For example, Van Dijk et al. (2010) reported that climate change, especially elevated temperatures, can change the abundance, seasonality, and spatial distribution of helminths (Nematoda / round worms and Trematoda / flatworms) that are parasitic to livestock. Changes in moisture and temperature conditions and growing season can potentially affect the growth of mycotoxins in grains, especially corn (Nardone, Ronchi, Lacetera, Ranieri, & Bernabucci, 2010). These myriad factors may in turn affect GHG emissions from the livestock system.

In general, dynamics of grassland ecosystems, such as those in the Great Plains, can be altered by changes in plant nutrient-use efficiency, water use efficiency, present plant species, biomass production, nutrient cycling, forage consumption by animals (livestock, wildlife, and insects), plant disease, and rate of biomass decomposition. Many of these can be potentially affected by climate change (King et al., 2004; Morgan, Derner, Milchunas, & Pendall, 2008).

Changes in temperature affect both the rates of chemical reactions and exchanges of energy between the land and the atmosphere. Kinetic responses have the potential to increase plant growth (Luo, Li, Jiang, & Polle, 2009), speed up plant development (Cleland, Chiariello, Loarie, Mooney, H, & Field, 2006; Hovenden et al., 2008; Sherry et al., 2007), and increase the decomposition of soil organic matter, although those potentials can be limited or altered by soil moisture. As a result, warming may increase the rangeland plant growth in mesic systems or during years with adequate moisture, but may have little effect (Fay et al., 2011; Morgan et al., 2011; Pendall, Osanai, Williams, & Hovenden, 2010; Xia, Niu, & Wan, 2009), or even reduce plant growth when soil moisture is inadequate and where warming leads to significant desiccation through increased evapotranspiration (De Boeck, Liberloo, Gielen, Nijs, & Ceulemans, 2008).

Climate change may affect precipitation patterns that will subsequently affect rangeland productivity (Lauenroth, Burke, & Paruelo, 2000; Sala, Parton, Joyce, & Lauenroth, 1988) and, ultimately, the carrying capacity of the range. However, more recent research suggests that response of grazing lands to precipitation depends not only on the annual amount, but also on frequency and size of precipitation events (Fay et al., 2011; Fay, Kaufman, Nippert, Carlisle, & Harper, 2008). Furthermore, differences in evapotranspiration, plant community and soil type can affect how precipitation variations affect soil water, plant utilization and species responses (Bates, Svejcar, Miller, & Angell, 2006; Craine, Spurr, McLauchlan, & Fierer, 2010; Debinski, Wickham, Kindscher, Caruthers, & Germino, 2010; Knapp et al., 2008; Whitford & Steinberger, 2011). An interesting example of this complexity is illustrated in a recent report in which less frequent precipitation events decreased aboveground net primary productivity (NPP) in tallgrass prairie, but increased NPP in shortgrass steppe (Heisler-White, Blair, Kelly, Harmoney, & Knapp, 2009). Thus, the specific effects of precipitation patterns can vary considerably across the region.

In addition to its effects on global warming, rising atmospheric levels of CO_2 affects plants directly as a substrate for photosynthesis and as an anti-transpirant. The former response is stronger in C_3 plants, like cool-season grasses, as their photosynthetic metabolism is not CO_2-saturated at present atmospheric levels; increases in CO_2 can potentially increase photosynthesis and plant growth in C_3 plants. That is not the case for C_4 plants (mostly warm-season grasses in rangelands) whose photosynthetic apparatus is CO_2-saturated or nearly so at present-day CO_2 concentrations (Anderson, Maherali, Johnson, Polley, & Jackson, 2001; Polley, 1997; Poorter & Navas, 2003; Reich et al., 2001). However, both herbaceous C_3 and C_4 plants experience the closure of leaf pores or stomates with rising CO_2 and the resultant decreased leaf transpiration (Wand, Midgley, Jones, & Curtis, 1999). As a result, rising CO_2 can significantly increase water use efficiency, especially in grasslands (Morgan et al., 2004; Polley, Jin, & Fay, 2011), so much that it may offset desiccation resulting from moderate levels of warming (Morgan et al., 2011).

Plant community composition largely governs important ecosystem attributes, such as net primary production, water and nutrient cycling, and plant-animal interactions. While shifts in plant community species composition in response to global changes are likely already underway, predicting particular species or functional group responses remains challenging (Polley et al., 2010). Vegetation shifts are expected to occur gradually, although abrupt changes due to crossing critical environmental thresholds are likely to happen as well (Craine et al., 2010; Fay et al., 2011; Friedel, 1991; Polley et al., 2011). Vegetation changes will sometimes involve complex interactions of one or more global change factor influencing the susceptibility of vegetation to disturbances, like fire (Bond, 2008). For example, in the Great Plains, the expansion of tree islands in native grasslands is likely due to fire removal, but may be exacerbated by rising CO_2 concentrations (Morgan et al., 2008).

The quality of grassland forage is an important determinant of livestock performance. While both rising CO_2 and temperature can reduce forage quality (Akin, Fales, Rigsby, & Snook, 1987; Craine et al., 2010; Gentile, Vanlauwe, & Six, 2011; Henderson & Robinson, 1982a, 1982b; Morgan et al., 2008; Newman, Sinclair, Blount, Lugo, & Valencia, 2007), complex interactions between global change factors and the environment suggest that both increases and decreases in forage quality are possible. Similarly, plant species shifts may also result in either increased (Polley et al., 2011) or decreased (Morgan et al., 2008; Morgan, Milchunas, Lecain, West, & Mosier, 2007) forage quality. The combined effects of climate change on species composition and nutrient cycling are likely to affect forage quality differently in different rangeland ecosystems, so that rangeland managers will need to carefully monitor their resources.

A decrease in livestock carrying capacity will occur in areas that receive less rainfall as they transition to shorter grasses and as biomass production decreases. In some cases, higher CO_2 concentrations may increase forage production but may decrease protein (and possibly other nutrients) in the forages (King et al., 2004; Milchunas et al., 2005). Additionally, severe droughts may remove forage and increase susceptibility to less palatable species. These changes in rangeland productivity may be partially mitigated by changes in supplementation strategies, changes in grazing management, transitions to mixed grazing (sheep and/or goats with cattle for example), or an increased use of wildlife as part of the ranching operation. Changes in precipitation may not only affect production of grassland monocultures, but may alter the predominant grasses and forbs present.

Changes in atmospheric CO_2 concentration, nitrogen deposition in soils, rainfall (total and patterns), and/or temperature all may affect plant productivity and/or biodiversity. However, increases in temperature may also increase the rate of biomass breakdown in soils and the release of stored C as CO_2. Under drier conditions, the short grass steppe may migrate eastward where climate conditions in the future will be similar to eastern Colorado historically. With or without this migration, productivity of the mixed- and tall-grass prairies will be reduced due to the weather patterns that are less conducive to the species present under current climate conditions. The effect would a decrease in the quantity of forage produced per acre, and, thus, a decline in the carrying capacity of these rangelands.

Climate change may also alter the suitability of land to grow crops or forage intended for livestock feed, particularly in drought prone areas. If the costs of moving feed, irrigation, or fertilizer usage increased substantially, it could result in a change in the location of intensive livestock production operations, such as feedlots, or alternatively, decreases in the number or size of feeding operations.

Climate change may also increase the frost-free period and subsequently alter the competitiveness of plant species, plant diseases, and pests. In 2000, the average growing season in the lower 48 US states was about 10 days longer than the 100-year average, due to a combination of later first frosts and earlier last frosts. Interestingly, the increase in the growing season for the past 30 years is almost a mirror image of shorter growing seasons which occurred in the early 20th century (U.S. Environmental Protection Agency, 2010).

Nardone et al. (2010) hypothesized that livestock systems based on grazing and mixed farming systems will be more affected by global climate change than more intensive confinement systems. The effects may differ by region but more intensive systems may be able to adapt more easily to changes than extensive systems. Although it has not been studied in depth, Nardone et al. (2010) suggested that climate change may affect the health of farm animals both directly and indirectly by the effects on disease vectors and/or on host resistance to disease. Adapting to the stressors of climate change may result in altered nutrient intake (via effects on feed intake and quality of forages) and decreased animal performance.

An increase in drought could lead to increases in rangeland and/or forest fires. NASA estimates that fires annually consume 1.8 billion to 10 billion metric tons of biomass and release billions of tons of GHG annually (Cawood, 2011). Scientists at the Australian Commonwealth Scientific and Industrial Research Organization estimated the GHG emissions from burning or feeding one ton of grass to cattle and found that GHG intensity of burning grass was approximately 3.6 times greater than if the grass was consumed by cattle (Cawood, 2011).

MITIGATION STRATEGIES: LIVESTOCK PRODUCTION

Moss et al. (2000) noted that world methane sources totaled about 759 million short tons/year (689 Tg/year) with an annual excess of about 93 million short tons (84 Tg). They suggested that atmospheric methane is increasing at a rate of about 33 to 44 million short tons (30 to 40 Tg) annually. Decreasing this trend would require reductions in methane generation and/or increases in methane sinks (such as oxidation in soils) (Ojima, Valentine, Mosier, Parton, & Schimel, 1993). They estimated that without temperate forest and grassland ecosystems the increase would be approximately 1.5 times the current rate. Moss et al. (2000) calculated that the reduction in methane generated annually, required to stabilize atmospheric methane concentration at current levels is approximately 10% of anthropogenic emissions.

Grasslands have the capacity to sequester carbon and to oxidize methane (Mosier, Delgado, Cochran, Valentine, & Parton, 1997; Mosier, Morgan, King, LeCain, & Milchunas, 2002; Mosier, Pendall, & Morgan, 2003; Ojima et al., 1993; Soussane, Tallec, & Blanford, 2010). However, carbon sequestration is both reversible and vulnerable to disturbance and climate change. A number of management practices are capable of affecting carbon sequestration including: 1) soil tillage and conversion of grasslands to crops; 2) moderately intensifying nutrient-poor permanent grasslands; 3) using light, rather than heavy, grazing; and 4) converting grassland to grass-legume mixtures (Ojima et al., 1993; Soussane et al., 2010).

Soil organic matter is generally greater in soils of the Northern Great Plains than the Southern Great Plains, suggesting that soil respiration and organic matter decomposition are greater in warmer areas than colder regions (Epstein, Burke, & Lauenroth, 2002). However, Epstein et al. (2002) reported that temperature accounted for less than 8% of the variation in organic matter decomposition rate and that increased soil moisture (> 30%) and decreased clay content were major drivers in soil organic matter content. Plant productivity declined with increasing temperature, suggesting that the lower soil organic matter in the south was not directly due to temperature, but indirectly to less biomass production.

An estimated 102,000 tons (93 Gg) of excess nitrogen are applied to cropland in the Great Plains annually (84,000 tons (76 Gg) in the North and 18,000 tons (16 Gg) in the South) compared to 298,000 tons (270 Gg) in the Corn Belt, 185,000 tons (169 Gg) in the Great Lake States, 36,000 tons (33 Gg) in Appalachia, and 44,000 tons (40 Gg) in the Northeast (Ribaudo, 2011). More efficient use of fertilizer nitrogen on crops and pasturelands could potentially decrease the cost of production and simultaneously decrease N_2O emissions (Liebig, Gross, Kronberg, Phillips, & Hanson, 2010).

Table 7.2 Effects of management strategies on GHG emissions

Scenario	Total GHG (T CO2e)	Total carcass wt(T)	GHG, CO2e/kg carcass	Change in GHG intensity from baseline, %
Baseline	5446	250.6	21.73	--
Feedlot:				
Increased forage use	5925	256.1	23.14	+ 6.59
Extended grain feeding	5277	247.2	21.35	-1.76
Feeding oilseeds to stockers	5371	250.6	21.43	-1.37
Feeding oilseed to finishers	5360	250.6	21.39	-1.57
Feeding DDG to stockers	5390	250.6	21.51	-1.02
Feeding DDG to finisher5404	5404	250.6	21.56	-0.77
Breeding stock:				
Feeding oilseeds	4986	250.6	19.89	-8.44
Feeding DDG[a]	5140	250.6	20.51	-5.62
Improved forage quality	5182	250.6	20.68	-4.85
Increased longevity	6191	286.2	21.63	-0.44
Increased calves weaned	5561	265.9	20.92	-3.74

[a] DDG= distillers grains. When dietary fat levels are held constant, the feeding of distillers grains will not affect enteric emissions or increase emissions (Hales et al., 2012: and unpublished data)

Source: (Beauchemin et al. 2011)

Beauchemin et al. (2011) estimated the effect of numerous strategies on the GHG emissions of a beef herd (cow-calf through finish) using the HOLOS model and noted the greatest possibility of reductions occur in the cow herd, rather than in the feedlot (Table 7.2). A number of practices, such as feeding of ionophores, supplemental fat, increasing dietary grain content, grinding forages, and increased grain processing have been shown to decrease enteric methane emissions from cattle. A number of other methods, such as feeding organic acids (fumarate, malate), probiotics, tannins, and saponin, have also been tested with mixed success, (K. Beauchemin, Janzen, Little, McAllister, & McGinn, 2010; Grainger & Beauchemin, 2011; Haaland, Matsushima, Johnson, & Ward,

1981; Hales, Cole, & MacDonald, 2012; Martin, Morgan, & Doreau, 2010). A number of these strategies are already used in many feedyards (Table 7.2)

ADAPTATION STRATEGIES: LIVESTOCK PRODUCTION

In the future, due to increases in world population, there will be increasing competition for land to produce food for people, bioenergy crops, and feed for livestock. Livestock producers may need to modify their nutritional and management strategies in order to compensate for changes in the quantity and quality of feed resources caused by climate change and competition for land. In addition, they may increasingly adopt mitigation strategies in order to decrease GHG emissions and/or earn carbon credits. Factors such as consumer beef demand and government policies and regulations could also affect the strategies adopted.

These modifications may include changes in the dominant species used on rangelands (cattle vs. sheep/goats vs. wildlife), increased use of mixed species grazing, changes in stocking rates, or changes in the phenotype and/or genotype of the animals used. For example, smaller cows with lower milk production have lower nutrient requirements than larger cows or cows with high milk production and thus require less forage and less supplemental feed. Selecting for smaller cows may be favorable in some regions for reducing GHG emissions and/or GHG intensity.

Environmental factors will potentially affect how cattle and calves move through the beef cattle production sectors. For example, in periods of drought, the quantity of forages available will be limited; thus, stocker calves may spend less time on pasture (and more time in the feedlot) and/or producers may sell portions of their cow herd in order to have sufficient forage for the remaining animals, but reducing breeding stock inventory during drought recovery years.

In some cases, higher CO_2 concentrations may increase forage production but may decrease protein (and possibly other nutrients) in the forages (King et al., 2004). Thus, changes in supplementation strategies (i.e., greater protein supplementation of cows or stockers), and grazing management may be required. Bailey (2004), Haan et al. (2010), and others have noted that grazing and supplement management can alter cattle distribution on pastures, the distribution of urine and feces on the pasture, and the efficiency of forage harvesting and utilization. By improving management strategies, grazing distribution and utilization of available forage may be improved.

A decrease in forage production could result in a decrease in cow numbers or movement of cows from one region to another, where more favorable weather conditions occur. The number of cows plus calves and stocker calves (i.e. carrying capacity) that can be maintained on different rangelands and pastures vary depending on the species of grasses, season of the year, size of cows/calves, and precipitation. A general rule of thumb is that cow-calf producers stock sufficient cows to consume 75 to 80% of forage available in a typical year. In such a case, the producer will retain some of his calves and/or purchase stocker calves to graze the remaining 20-25% of forage. In wetter than normal years, they may purchase more stocker calves and in drier years will purchase fewer (or no) stocker calves. Thus, in drier than normal years, calves may enter feedlots at an earlier age due to a shortage of forage.

Using an economic model, Torell et al. (2010) calculated that under relatively constant environmental conditions the optimal cow-to-stocker ratio on native range in eastern New Mexico was about 80:20 – the ratio typically seen in much of the Great Plains. However, under highly variable conditions the optimal ratio was about 50:50. In the same vein, Okayasu et al. (2011) discussed differences between equilibrium and non-equilibrium environments in pastoral systems. In equilibrium environments where rainfall is relatively stable, the ratio of grazing animals to vegetation are "density dependent," and thus it is appropriate to calculate average carrying capacities and to use them to define sustainable animal populations. In contrast, non-equilibrium environments are characterized by large fluctuations in factors, such as rainfall and forage production, and thus in the carrying capacity of the rangeland. Under non-equilibrium environments, livestock producers have to adapt by moving animals between pastures with better conditions. They suggested it is important to identify and monitor boundaries between equilibrium and non-equilibrium environments, so that managers can respond to climate change. Similarly, using their economic model, Torell et al. (2010) noted that for optimal economic returns, producers need to have a flexible grazing management system, where livestock numbers are adjusted to match the available forage.

Intensive livestock production systems will also have to adapt to climate change. Because of social issues, geography, topography, nutrient management, and environmental constraints, the Great Plains will remain a major cattle feeding area, although some movement from the Southern Plains to the Northern Plains, where bio-ethanol and corn starch industry byproducts are more available and feed costs are lower, may occur. In addition, many climate change projections estimate the Southern Plains may be more negatively affected than the Northern Plains. Increased use of grains and forages for bioenergy and/or human food will limit the quantity of feed grains available, and will result in increased use of byproducts and other feeds ingredients unacceptable for use in bioenergy or human consumption.

Changing Socio-Economic Factors Influencing Agriculture

Over decades, US agricultural producers have faced shrinking profit margins and received a reduced portion of agricultural profits at the farm-gate. These trends have been related to a wide range of economic, energy, and agricultural policies that have led to globalized markets, aggressive competition in international trade, and rapidly evolving high-capital agricultural technologies. Other economic policies have led to a higher portion of production coming from larger farms with a decline in mid-sized farms. While the larger farms may have capacity to adopt more efficient production practices and systems, small farms continue to be important, making up 88 percent of US farms in 2007, holding 63% of agricultural land, and marketing 16% of farm product. Small farms accounted for 76% of land enrolled in USDA land-retirement programs, indicating their significant role in natural resource and environmental outcomes of agriculture within landscapes (Hoppe & Banker, 2010). Urban agriculture provides potential for higher income from small, fragmented landscapes in the rural-urban fringe, and helps maintain an abundance of environmental services, including hydrologic function, which may have increasing importance in an era of urban heat islands and climate change.

Increased farm size, reduced farm numbers, and reduced population have greatly decreased the capacity of many Great Plains rural communities to support agricultural, economic, and social infrastructure. However, most Great Plains residents reside in urban areas or in metropolitan-influenced counties (Parton, Myron, & Ojima, 2007). Emerging local-foods marketing opportunities provide potential for young farmers and more diverse farmers to get a start in agriculture. The US local food market was $4.8 billion in 2008, and small farmers utilized local markets at a higher rate than larger farms (Low & Vogel, 2011). Small and midsized farms with local food sales were more likely to list farming as the principal operator's primary occupation than small farms that did not utilize local sales. While the potential for small farms and local food markets is obvious for the more urban portions of the Great Plains, many of the US's most food-insecure counties are in the rural Great Plains, where families are long distances from grocery stores and households lack access to cars, due to health or poverty (Ver Ploeg et al., 2009). The need is great for farmers markets, community supported agriculture, community gardens and other local food enterprises in the rural Great Plains.

Great Plains agriculture has been dramatically shaped by increasing energy costs, including the influence of energy prices on fertilizer, equipment, and on-farm fuel use. Intensive production systems that were developed in an era of relatively inexpensive energy, such as irrigated agriculture and large confined animal feeding operations, face many challenges to maintain profitability under the new economic conditions.

Upwards of 70 percent of the Great Plains region is classified as range and cropland, producing a variety of crops and livestock. While the total land cover devoted to agriculture has remained relatively unchanged over the last few decades, the crop mix within the region has changed as economic, social, environmental, and technological variables have shifted. The mitigation of greenhouse gases and the subsequent push for biofuels is one such development that has had large impacts on land-use change in the region given its spatial extent, or what has been referred to as "energy sprawl" (McDonald, Fargione, Kiesecker, Miller, & Powell, 2009). For instance, acres devoted to producing corn, a major bioenergy feedstock, have increased by roughly 32 percent between 1997 and 2007, or by over 5 million new acres (2 million hectares). This change in crop mixture has and will continue to impact the demand for major inputs of production, especially water. However, biofuel-driven land-use changes also have indirect effects on GHG emissions that may offset some of its benefits, such as changes to the surface energy and water balance from landscape modification, which need to be considered to ensure that emissions have a net decline (Georgescu, Lobell, & Field, 2011).

New bioenergy markets provide great opportunity for agriculture, but also present societal challenges associated with the potential competition between food, energy, soil and water conservation, and greenhouse gas mitigation needs. Bioenergy is the use of various forms of biological material that is grown and produced either directly for energy (e.g., corn for corn ethanol) or in the form of second-generation biomass or waste (e.g., agricultural waste, forest industry waste, municipal paper and wood waste). Bioenergy is seen as an area of potential economic development in the United States, especially in rural areas, as well as a potential source to contribute to domestic energy independence and reduction of fossil fuel use and greenhouse gas emissions (Pate, 2011). Adler et al. (2007) evaluated several bioenergy crops proposed for expansion

Table 7.3 Embodied Water for Ethanol (EWe) and Total Consumptive Water (TCW) in ethanol producing states in 2007. All numbers listed are in million gallons (1000 m^3)

State	Ethanol Production	EWe	Ground-water	Surface Water	Wir*	Wp*	TCW	Corn processed into ethanol
North Dakota	133 (505)	16 (59)	8 (31)	7 (28)	7,435 (28,146)	482 (1824)	7,917 (29,970)	18%
South Dakota	582 (2203)	25 (96)	10 (38)	15 (58)	53,828 (203,762)	2,100 (7950)	5,736 (21,712)	39%
Nebraska	655 (2481)	132 (501)	111 (422)	21 (80)	326,286 (1,235,128)	2,365 (8954)	328,652 (1,244,082)	16%
Kansas	212 (804)	139 (528)	128 (486)	11 (42)	111,438 (421,840)	767 (2903)	112,205 (424,743)	15%
Colorado	85 (322)	311 (1176)	60 (226)	251 (950)	99,615 (377,082)	307 (1161)	99,921 (378,243)	20%
Wyoming	5 (19)	358 (1354)	33 (125)	325 (1229)	6,749 (25,547)	18 (68)	6,767 (25,615)	23%

*Wir = irrigated water; Wp =process water
Source: (Chiu et al. 2009)

and found that switchgrass (one of the species most often proposed for Great Plains cellulosic energy production) would reduce GHG emissions by about 115% compared with the life cycle of gasoline and diesel, ethanol and biodiesel from corn rotations. The potential of bioenergy is entirely dependent on the form of biomass used and the variation of local and regional practices and conditions (Pate, 2011). For example, there has been a transition from soybeans to corn to produce ethanol and, in irrigated areas of the Great Plains, corn requires more water for irrigation than soybean or other crops that may have been displaced by corn. (Tidwell, Cha-tien Sun, & Malczynski, 2011).

There will also be regional variation in terms of the water-energy nexus, as research to date shows that the amount of water needed for biomass production can vary significantly across the United States as a whole and within the Great Plains region. For example, Table 7.3 below shows the embodied water for corn ethanol (EWe) production and total consumptive water (TCW) in states across the Great Plains (Chiu, Walseth, & Suh, 2009). This study shows that across the United States water requirement estimates -- from corn at the farm to fuel at the pump -- range from 1.3 to 565 gallons (5 to 2138 liters) of water per liter of ethanol (EWe), and, in the Great Plains, the estimates range from a low of 16 million gallons (59 million liters) EWe in North Dakota to 358 million gallons (1354 million liters) EWe in Wyoming (see Table 7.3 below). However, it is worth noting that a high EWe does not necessarily translate directly into a high TCW as in the case of Wyoming. This highlights the need to understand local and regional specifics in terms of conditions and practices when considering the potential and water-energy nexus of biofuels (Chiu et al., 2009).

In the Missouri River Basin, current assessments indicate that significant water demands will arise from requirements to meet the biomass production to support the Renewable Fuel Standard goals in 2030, even without the inclusion of additional climate change effects (Foti, Ramierz, & Brown, 2011). The agricultural demand in 2030 in the Missouri River Basin is projected to increase by about 30 million gallons per day (mgd) (113,600 cubic meters per day) due to industrial and urban consumption. Agriculture would increase by around 158 mgd (598,095 cubic meters per day) to meet the Renewable Fuel Standard goals (Foti et al., 2011).

One limitation on the expansion of corn ethanol production in the Great Plains is the use of groundwater in already vulnerable and water-stressed areas. For example, the TCW for the Great Plains states in 2007 amounted to 85 million cubic feet/ 2.4 trillion liters and 160 million cubic feet/ 4.5 trillion liters in 2008 (Chiu et al., 2009). In 2007, 68% of this water was supplied from groundwater in the already vulnerable Ogallala Aquifer region and, in 2008, the amount of water extracted accounted for approximately 18% of the entire annual rate of aquifer depletion (Chiu et al., 2009). One estimate found that in Nebraska and Kansas 15-19% of irrigation water went to growing corn for ethanol (Mishra & Yeh, 2011). Careful consideration must be given to producing corn ethanol in areas that are not already at high risk for water stress. A 2003 GAO report named several states in the Great Plains region as being threatened by water shortages across local, state, and regional scales (General Accounting Office, 2003). Yet, there is an economic incentive and strong pressure to grow corn for energy in the High Plains, where irrigation costs are only 20% of total production costs for corn, yet yield for energy crops can be increased significantly through irrigation (Tidwell et al., 2011). Experts predict that the nation's highest competition for water among biofuels and other demands will be in the High Plains region (Tidwell et al., 2011). This is one example of the tradeoffs that must be considered between fossil fuel energy, renewable energy to meet renewable portfolio standards, and the water necessary to meet these economic and environmental goals. It is critical to take regional and local context into account for policy and planning across all scales of governance; and it requires careful planning at the watershed level within and between states at a regional level (Pate, 2011). Some experts have suggested "next generation" biofeed stock, such as perennial grasses and woody biomass, will help meet the needs for bioenergy, however, the extent to which this potential exists or is limited by local and regional conditions in the Great Plains has yet to be determined (Christensen et al., 2007).

Due in large part to government mandates, production of ethanol from feed grains has increased exponentially over the past 12 years, from approximately 54 plants producing 1.7 billion gallons (6.4 billion liters) of ethanol to over 200 plants producing 13.5 billion gallons (51.1 billion liters) of ethanol. In the Great Plains states, there are approximately 69 bioethanol plants with capacity to produce about 4,365 million gallons

Table 7.4 Bioethanol plants in the Great Plains and border states

State	No. of plants	Total capacity, million gallons of ethanol	Total capacity, million cubic meters of ethanol
Colorado	4	125	0.5
Kansas	12	520	2.0
Montana	0	0	0
Nebraska	26	1,693	6.4
North Dakota	6	594	2.2
Oklahoma	0	0	0
South Dakota	15	1,066	4.0
Texas	4	355	1.3
Wyoming	2	12	0.05
Border States			
New Mexico	1	54	0.2
Arkansas	0	0	0
Missouri	5	251	1.0
Iowa	39	3,370	12.8
Minnesota	22	1,331	5.0

Source: (RFA, 2011)

(16.5 million cubic meters) of ethanol annually (Table 7.4). The "border states" have an additional 67 bioethanol plants that have the capacity to produce 5,000 million gallons (19 million cubic meters) of ethanol annually. In the US, there are about 200 ethanol plants using about 30% of the U.S. corn crop. However, recent policy changes, which eliminated some of the subsidies related to corn ethanol production, may alter the usage of corn products to produce ethanol in the region.

Distiller's grains are a byproduct of the grain ethanol industry. Each bushel of corn, 56 lb (25.4 kg) produces about 18.7 lb or 2.83 gallons (8.5 kg) of ethanol, 18.7 lbs (8.5 kg) of CO_2 and 18.7 lbs (8.5 kg) of dried distiller's grains. Corn gluten feed is a byproduct of the corn starch/sweetener industry. Over 35.3 million short tons (32 million metric tons) of distiller's grains and 5.5 million short tons (5 million metric tons) of corn gluten feed were produced in 2010. The beef cattle industry consumes approximately 41% and the dairy industry consumes approximately 39% of all US distiller's grains produced (Renewable Fuels Association, 2011). Approximately 61% was fed in the dry form and 39% in the wet form. Feeding these byproducts to cattle in the wet form has several advantages over feeding the dry product: most notably, it saves the high cost of drying material which contains about 70% moisture down to 10% moisture and avoids spoilage.

Today at least 30% of all US corn production is used in the bioethanol and corn sweetener industries. Government ethanol policies and other factors will determine if this trend continues. The primary use of these byproducts will probably continue to be livestock feed. However, changes in production procedures (for example, removing the fat from distiller's grain for use as biodiesel) may alter the feeding quality and demand for these byproducts.

© Danny Martin www.dannymartinphotography.com

Chapter 8

Great Plains Societal Considerations

Impacts and Consequences, Vulnerability and Risk, Adaptive Capacity, Response Options

A variety of factors related to climate variability and change will impact the Great Plains across human and ecological communities. The changes and associated stress are triggering response strategies and other mitigation and adaptation measures from land managers, government officials and staff, and various industries. The impacts and responses address water, energy, and other essential resources for both human and environmental well-being.

Based on modeled projections of climate change, scientists, land managers, city-managers and others are already implementing mitigation and adaptation strategies for agriculture and livestock production and other elements of the regional economy. Response strategies consider ecosystem services that benefit Great Plains communities and the biological and ecological changes that may affect wildlife and their habitats, including wetlands and river systems.

Low-income communities, including Native American reservations and colonias along the US-Mexico border, are among the most vulnerable to climate change effects in the Great Plains. In many of these places as well as in cities and urban regions, managers and businesses are establishing pilot projects to adapt more resilient resource uses and construction practices.

Trends and models also suggest changes in regional climate and the frequency and severity of extreme weather events. Additional impacts include shifts in disease distributions, representing health risks through potential outbreaks. These factors have led the insurance industries to reconsider the elevated economic and human risks and vulnerabilities, complementing scientific research of ongoing and projected climate change.

Urban-Rural Dynamics

CLIMATE CHANGE IMPACTS ON URBAN AREAS

Urban areas currently face a wide variety of environmental challenges, many of which may be exacerbated by climate change. One such issue is ground-level ozone. Ground-level ozone is a known pulmonary irritant and is the primary constituent of smog (Ebi and Mcgregor 2008). A number of Great Plains urban areas have issues with ozone compliance, but the Houston area in particular, has been in non-attainment of the EPA ozone standard since it was set in 1977 (Raun 2010). Higher temperatures may result in greater ozone formation because the chemical reactions resulting in ozone formation are temperature dependent (Bell et al. 2005a). In addition, biogenic volatile organic compounds, which are ozone precursors, increase as temperatures rise (Bell et al. 2005b).

Suspended particulate matter also presents a potential air quality issue in cities. Sources of particulate matter (PM) include construction sites, smokestacks, fires, emissions from power plants, and automobiles (U.S. Environmental Protection Agency 2012). PMs can penetrate deep into the lungs and cause health problems. Prolonged or severe droughts may result in dusty conditions and wildfires that can cause an increase in suspended particulates including smoke, pollen, and fluorocarbons (Centers for Disease Control and Prevention et al. 2010).

The urban heat island (UHI) effect occurs when cities have warmer air and surface temperatures than surrounding rural areas, particularly at night (Grimm et al. 2008, U.S. Environmental Protection Agency 2008a). This is attributable to a variety of causes, including decreased vegetation, lower albedo from impervious surfaces, and urban building morphology (Grimm et al. 2008, U.S. Environmental Protection Agency 2008a). In urban areas with 1 million or more people, annual average air temperatures can be 2-5 °F (1-3 °C) higher than surrounding areas. On individual clear, calm nights the UHI can be as much as 22 °F (12 °C) warmer. Smaller cities and towns can create heat islands as well, although the urban-rural temperature differences often decrease as the city size decreases (U.S. Environmental Protection Agency 2008a). UHIs can act in conjunction with climate change to create more extreme temperatures.

An increase in high temperatures, particularly long stretches of days over 100 °F, will damage the integrity of transportation systems. High temperatures, particularly those exceeding 90 °F (32 °C), can cause pavements to degrade faster, compromising their integrity (Savonis et al. 2008, Bjune et al. 2009). Increased temperatures can also cause some types of rail to develop "sun kinks" in which sections of the rail buckle (Savonis et al. 2008). Increased cooling and thus energy consumption may be required for freight and passenger operations (Savonis et al. 2008). Compounding the problem, crews responsible for construction and maintenance may not be able to work during times of extreme heat (Savonis et al. 2008, Bjune et al. 2009).

Extreme rain events could result in increased flooding if flows start to exceed the design capacity of a city's culverts and storm sewer system (Savonis et al. 2008, Bjune et al. 2009). Bjune et al. (2009) assess that this would present a problem for cities lying on flat terrain, as is the case with many Great Plains metro areas. More intense storms will also reduce clearance under bridges and increase erosion of road bases and bridge supports (Savonis et al. 2008, Bjune et al. 2009).

Climate change could also have a variety of impacts on municipal water supplies. The headwaters of many Great Plains rivers are in the Rocky Mountains, and cities in the western part of the region, such as Denver, are often dependent on snowmelt. The snowpack acts as a natural and massive reservoir for water storage, holding water historically until late spring or early summer. Warming temperatures will not only result in a decreased amount of snow and reduced water storage in the snowpack, but it will also cause snow to melt earlier in the spring (Barnett et al. 2005). In the absence of precipitation changes, maximum runoff will shift to earlier in the season, further from the peak water demand months of July and August.

In addition to shifting times of peak runoff, warmer temperatures may also affect evaporation rates. Many cities in the Great Plains are dependent on reservoirs for their water supplies, and these reservoirs currently lose considerable amounts of water to evaporation. Annual evaporation from the six largest reservoirs on the Missouri River's main stem, for instance, has been estimated to be about 5% of the average annual river discharge (Benke and Cushing 2005). In the Rio Grande, evaporation from the major reservoirs has been estimated to exceed municipal water usage in the basin (Benke and Cushing 2005). Such reservoir losses could increase if warmer temperatures persist without an increase in precipitation.

Increases in precipitation intensity could adversely affect municipal water supplies by causing elevated levels of turbidity, organic matter, pathogens, and pesticides in source waters, associated with either rises in nonpoint source pollution loads or increased infiltration influencing groundwater quality (Kundzewicz et al. 2008, Clark et al. 2011). For cities, such as Kansas City, in which storm and wastewater sewers are combined, high rainfall events could also overload the capacity of wastewater treatment plants leading to situations in which untreated or partially treated sewage may be discharged into streams (Kundzewicz et al. 2008, Delpa et al. 2009, Struck et al. 2009).

Droughts can lead to water-quality problems for municipalities, as well as water-quantity issues. In some areas, droughts may result in elevated levels of toxic algae and organic matter in source waters, and lower streamflows may lead to the concentration of pollutants. Such factors may adversely affect the ability of treatment plants to meet safe drinking water standards (Centers for Disease Control and Prevention et al. 2010). Excessive drying of soils can damage pipes leading to breaks in water mains, such as those experienced in Texas during the state's most severe one-year drought on record. In Houston alone, over 6000 water main breaks were reported during summer 2011 (Climatologist 2011, Houston City Council 2011, Royal Academy of Engineering 2011).

Green infrastructure is one approach that cities can use to simultaneously address these issues as they upgrade aging, outdated infrastructure. Although the term green infrastructure can have alternative meanings in different contexts, it often refers to landscapes that have been specially conserved or sometimes designed and engineered to mimic natural processes and provide ecosystem services, such as flood control (U.S. Environmental Protection Agency 2010, Foster et al. 2011). Sometimes the definition of green infrastructure is expanded to include additional approaches (not always vegetation-related) that cities use to try and achieve environmental goals (Foster et al. 2011). In the text below, it is this broader definition that is considered.

Green infrastructure can benefit climate change adaptation strategies through its ability to curb the impacts associated with the anticipated increases in air temperatures and in extreme precipitation events (Foster et al. 2011). Benefits associated with climate change mitigation are generally related to the ability of green infrastructure to decrease energy usage and sequester carbon. In addition, green infrastructure can also contribute to recreational space and aesthetic value that can improve health and provide a better quality of life (Tzoulas et al. 2007, Foster et al. 2011).

Green infrastructure approaches can be incorporated into new developments, completed as a retrofit or included as repairs or replacements are made. They can be implemented at an assortment of spatial scales ranging from individual house lots to entire metropolitan regions (Foster et al. 2011). Although green infrastructure may be implemented to meet a single, specific goal, such as reducing ambient air temperatures, it often provides additional benefits, and the full value of a project stems from the multiple functions that green infrastructure performs. A variety of cities within the Great Plains are beginning to incorporate green infrastructure into existing building codes and city plans as a pragmatic way to update current infrastructure to meet climate challenges.

Examples of initiatives being taken include: eco-roofs, such as cool roofs or white roofs (U.S. Environmental Protection Agency 2008b, Energy 2011, Foster et al. 2011); bioretention, to address flood control and water quality protection by creating vegetated depressions that receive, absorb, and treat stormwater runoff from impervious surfaces (U.S. Environmental Protection Agency 2010, Foster et al. 2011) to capture and remove contaminants and sedimentation; and urban forestry or greenways which can be developed to sequester greenhouse gases, providing natural cooling to buildings and pavement, improving air quality, reducing energy bills, decreasing stormwater runoff, controlling erosion, and adding attractive landscapes (Mid-America Regional Council n.d., Briechle 2009, Universitiy of Nebraska 2011, City of Grand Forks 2012, Denver Mile High Million Initiative 2012).

GREAT PLAINS URBAN WATER SUPPLY STRATEGIES

In the face of both population growth and greater uncertainty in precipitation and runoff regimes stemming from climate change, cities throughout the Great Plains are starting to explore and implement ways to diversify their water sources. Strategies include water conservation, the use of nonpotable water, aquifer storage and recovery, desalination, and water reuse, with the latter approach being the subject of more in-depth discussion in this Chapter.

Water conservation is becoming a priority throughout Great Plains cities. Cities, including Austin, Dallas, Denver, and San Antonio, all have water conservation plans or programs. The resulting decrease in demand can act as an "effective" new water source. Components of the plans vary and include indoor residential, commercial, and industrial approaches as well as outdoor conservation approaches. An example of an indoor residential conservation measure is Dallas Water Utilities', "New Throne for Your Home" program that provides vouchers to replace older, pre-1992 toilets with newer, more efficient models (City of Dallas 2010). An example of an outdoor conservation measure is Denver Water's soil amendment program that requires property owners to till compost into their soil before Denver Water will set meters so that the soil will retain water more efficiently, reducing irrigation requirements (Denver Water 2011a).

The use of nonpotable water in situations when water of drinking-water caliber is not required is a strategy being implemented in Norman, Oklahoma. The city, for example, is using wells not suitable for drinking water to help irrigate the Westwood Golf Course and the Griffin park complex (City of Norman, 2011).

Aquifer storage and recovery involves the injection of water into a well when water is available for storage underground. When needed, the water is then recovered from the

BOX 8.1

Recovery from Disaster: Greensburg, Kansas, 2007

Greensburg, Kansas, serves as an example of a town that embraced sustainability and used a tragedy as an opportunity to rebuild in a greener manner. Prior to May 4, 2007, it was a rural town similar to other Great Plains farming communities. The energy structure of the town, developed in the 1960s, was similar to many rural towns in the Great Plains, with electricity created largely from coal-based sources. On the evening of May 4, 2007, an EF-5, 1.7-mile- (2.7-km) wide tornado with wind speeds over 200 mph (320 km per hour) hit the town, destroying or severely damaging 90% of its structures and killing 11 people.

In the aftermath of the storm, community citizens resolved to rebuild a town that is prepared to face 21st century challenges. Key city leaders expressed interest in rebuilding a model green community, which generated enthusiasm among residents eager to demonstrate that challenges present opportunities and a disaster can be turned into a chance to foster resilience. The Department of Energy and other key organizations, including the National Renewable Energy Laboratory, quickly aligned their support and interest in helping Greensburg rebuild and demonstrate energy solutions that could be replicated in other communities. Other federal and state agencies, nonprofit, professional organizations and individuals reached out to Greensburg with professional expertise and donations of materials or cash.

On August 15, 2007, the City of Greensburg adopted a Long-Term Community Recovery Plan that was prepared through FEMA's Long-Term Community Recovery program, which included strategies to rebuild sustainably. The residents then developed a Sustainable Comprehensive Master Plan for the town's next 20 years. It states, "A truly sustainable community is one that balances the economic, ecological, and social impacts of development." In implementing the recovery plan, Greensburg has set a new standard for other rural and urban communities. It has become a net-zero energy community, generating as much electricity from renewable sources as it consumes. The city council passed a resolution requiring all new city buildings larger than 4,000 square feet (370 square meters) to reduce energy consumption by 42% (compared to standard buildings) and pass US Green Building Council LEED Platinum certification. An 11,000-BTU per second (12-megawatt) wind energy system will be installed near Greensburg that will meet its pre-tornado electricity needs. Additionally, the city has entered into a power purchase agreement from a renewable energy provider that will deliver 100% renewable electricity from wind, hydro, and other renewable energy electricity generation sources.

Greensburg citizens acknowledge that there is potential for similar disasters in the future and have adopted building code standards to be better prepared for severe wind events. It has also embraced tornado preparedness education within the community, and schools have implemented programs to educate students about storm safety and sustainable living. (City of Greensburg, 2008; National Renewable Energy Laboratory, 2009)

same well (National Research Council 2008). If water is recovered via a different well, the process is called aquifer storage transfer and recovery. The cities of El Paso, Kerrville, and San Antonio in Texas all have aquifer storage and recovery programs making use of treated wastewater, treated river water, and groundwater for injection, respectively (Texas Water Development Board 2011). San Antonio, for instance, pumps water from the Edwards Aquifer during wet periods and stores it underground in the Carrizo Aquifer. During times of drought, the stored water is then recovered to help meet peak water demands (San Antonio Water System 2009).

Advantages of water reuse include improved water supply reliability, in particular during droughts, and reduced dependence on imported water supplies. In some instances, reuse may increase the amount of water for the environment; for example, if it replaces some existing surface or groundwater supplies, thereby increasing instream flows or decreasing groundwater pumping. Water reuse may also improve surface water quality when nutrient-laden effluent is diverted for the irrigation of landscapes and crops (National Research Council 2011).

At the same time, water reuse could also potentially have negative effects on downstream flows and water quality. Depending on its extent and context, reuse may decrease downstream flows, adversely affecting downstream users and ecosystems, particularly in water-limited environments. If irrigation application rates exceed the ability for plants to make use of the nutrients in the reclaimed water, this could result in excess nutrient levels in ground- or surface water, which could lead to human health and environmental effects (National Research Council 2011). Irrigation with reclaimed water could possibly produce excess levels of salinity in soils, which can be detrimental to plant growth. Denver Water has been studying this and exploring options for decreasing impacts (Denver Water 2011b). Depending on project design and energy sources, reuse projects also have the potential to increase the carbon footprint of water supplies (National Research Council 2011).

The financial costs of water reuse projects vary and are highly site specific (National Research Council 2011). They depend on a variety of factors including the degree, if any, of additional treatment needed before reuse, pumping requirements, timing and storage requirements, and the extent of any new transmission pipelines. This latter factor is related to the distance between a wastewater treatment plant and reclamation plant, the need for and sizing of any piping for the conveyance of nonpotable water, which has to be kept separate from the potable transmission lines already in place, and the distances between the reclamation plant and non-potable water customers (National Research Council 2011). In combination with water conservation, water reuse could potentially decrease seasonal peak demands, which can reduce capital and operating costs (National Research Council 2011).

A 2011 National Research Council report on US water reuse notes that if utilities decide to start placing more emphasis on water reuse, moving towards having multiple smaller, decentralized wastewater treatment plants could make more sense. Currently, wastewater treatment plants are generally constructed at low elevations near a discharge point such as a river or lake. Consequently, reclaimed water must generally be pumped uphill for use. A more decentralized system in which reclaimed water is closer to potential customers could reduce pumping costs as well as the costs of transmission and distribution infrastructure. In addition, such a system might be able to better accommodate demand fluctuations in contrast to a large, centralized plant.

The 2011 National Research Council report on water reuse in the US also notes a variety of research needs. Included among these are conducting an analysis of the extent of de facto potable water reuse in the US and improving our understanding of the health impacts of human exposure to constituents in recycled water. The report also notes that while water reuse for ecological enhancement is promising, few studies have examined possible environmental risks.

Rural and Tribal Landscapes: Contrast and Comparison Vulnerability, Opportunity, and Adaptive Capacity

CLIMATE CHANGE IMPACTS ON RURAL AREAS

The impacts of climate change on rural communities are determined by a set of complex interactions among the environment, different sectors, and population groups (Parton et al. 2005, 2007, Hartman et al. 2011). The potential impacts of climate change include the modified vulnerability of rural families dependent on farm and ranching activities to climate and market stresses; the modification of crop and livestock production systems; water use competition; changed water quality; expansion of weeds, pests, and diseases; a change in plant-animal communities; altered fire and storm patterns; changes in grassland ecosystems and species composition; disruption of pollinator relationships; tree mortality; enhanced vulnerability to drought conditions, and insect or disease outbreaks in a number of ecosystems (Parmesan and Yohe 2003, Parmesan 2006, Field et al. 2007, U.S. Climate Change Science Program 2008a, 2008b). There is a scarcity of information and literature on the interface of how socioeconomic and demographic factors will interact with the biophysical changes accompanying global change and almost no information on how the interconnected socio-economic / ecological systems will respond (Lal et al. 2011).

One certainty is that vulnerability to climate change is intensified in rural areas with highly climate-sensitive livelihoods, where communities have fewer resources and alternatives than metro areas. Lal et al. (2011) suggest that rural areas typically have higher poverty rates and lower household incomes, historically putting them at higher climate-related risk from weather-related shocks. The impacts of climate change and capacity to manage resulting challenges will vary across the region and within communities, just as households have differentiated vulnerabilities and coping mechanisms. A range of impacts will be felt across different communities, with some benefiting from climate-induced changes, and other facing devastating losses. Further regional research that improves upon current understanding of socio-economic and biophysical impacts of global change on rural communities would be useful to develop appropriate policies and mitigate negative consequences (Lal et al. 2011).

As stated in Chapter 7, the response of agricultural systems to climate change will vary across the region. However, the disproportionate percentage of rural counties (versus metro counties) reliant on agriculture as a primary source of economic activity suggests rural communities will experience the brunt of climate impacts on agriculture

(Lal et al. 2011, USDA Economic Research Service 2012). If yields decrease, not only will profits and income be lowered, but families reliant on agriculture for subsistence will be doubly impacted by both a loss of income and food source. Similarly, farming communities are expected to experience additional water stress from climate change, particularly in counties reliant on irrigation. Chapter 4 details critical issues related to the effects climate change will have on water. Aquifers in the Great Plains continue to be tapped faster than the recharge rates, causing unsustainable water-use in the region (Barnett et al. 2008). Although urban areas are using more total water, the greatest percentage is surface-water. On average, rural communities (including agriculture) use more groundwater and almost eight times the total water of urban areas (U.S. Geological Survey 2005).

Effects of climate-related events on social systems are less known but can be expected to be negative in remote areas. As previously stated, the accessibility of health care resources tends to deteriorate as population density declines. With decreased access to health infrastructure and a higher proportion of income spent on health services, rural communities are likely to become more vulnerable to the harmful climate change health impacts discussed later in this chapter.

The Native American people of the Great Plains have lived in this region for thousands of years. However, as the region deals with the current challenges of the 21st century, the added stress of climate change on socioeconomic and political factors of the Great Plains is further exacerbating the degrading conditions of many of these tribal communities. How climate change impacts tribes in the Great Plains, with changing water conditions, health implications, and energy challenges are a concern throughout the region. How tribes in this region can draw on their cultural values in developing strategies that can be used to adapt to and mitigate climate change are lessons which can be shared across the rural communities in the region.

A number of tribal communities living in the rural areas have limited capacities to respond to climate change. Many reservations already face severe problems with water quantity and quality – problems likely to be exacerbated by climate change and other human-induced stresses. However, a number of communities and tribal governments are establishing strategies to cope with these social-ecological challenges related to environmental and climate changes taking place on their lands. These activities recognize the socioeconomic challenges faced by these communities, isolated areas where housing often lacks electricity and running water. Communities dealing with high poverty rates and poor health levels are indicators of communities more at risk to climate change. Native American populations on rural tribal lands have limited capacities to respond to climate change and ability to move is constrained by cultural and other socio-economic linked to the tribal lands.

Tribes are disproportionately impacted by rapidly changing climates, manifested in ecological shifts and extreme weather events, as compared to the general population, due to the often marginal nature and/or location of many tribal lands. The high dependence of tribes upon their lands and natural resources to sustain their economic, cultural, and spiritual practices, the relatively poor state of their infrastructure, and the need for financial and technical resources to recover from such events all contribute to the disproportionate impact on tribes (Intertribal Climate Change Working Group 2009). Tribal communities are deeply connected to local ecosystems and are economically and culturally dependent on the fish, wildlife, plants, and other resources of their lands. However, this connection to the local ecosystems and ecosystem services also provides potential long-term solutions for adaptation as these strategies incorporate ecosystem services as part of these actions to deal with climate changes in their social-ecological system. So there are ways which the various Indian tribes have shown significant strengths and resiliency to meet these challenges.

WATER

Water is vital for drinking, agriculture, economic activities, and ecological habitats – basically, for life. And while tribes have adapted to the water cycles of the Great Plains over generations, population growth, region-wide increased industrialization, and climate change are making the variable water supply and regimes in this area more uncertain. Tribes already face significant challenges in providing adequate water supply and wastewater treatment for their communities. Climate change will add to these challenges.

In addition, the uncertainty associated with undefined tribal water rights results in constraints in developing strategies to deal with water resource issues. These water right issues are made even more complicated by the fact that these are often cross-jurisdictional, cutting across intersecting tribal, municipal, state, and federal boundaries. Various court cases have attempted to resolve these issues (e.g., Winters vs. US 1908, Arizona vs. California 1963), however, questions still are unsettled. In many areas of tribal lands, water infrastructure is in disrepair or lacking (U.S. Environmental Protection Agency 2011). According to a 2007 Indian Health Services Report, approximately 40,000 tribal homes in the Great Plains region had water supply deficiencies and 24,000 had deficiencies related to wastewater treatment (Rogers 2007). Roughly 9,700 homes completely lacked either a safe water supply system or a sewage disposal system or both (Rogers 2007). These conditions lead to increased vulnerability to climate extremes, and emergency fixes may take time to implement and can be costly. For instance, during a 2003 drought in the Missouri River Basin, Lake Oahe levels dropped so low that silt and sludge clogged the sole intake pipe at Fort Yates, North Dakota, cutting off the water supply for residents of the Standing Rock Sioux Tribe for several days and causing an Indian Health Services hospital to be temporarily shut down. A temporary intake system was installed at a cost of about $3 million (Albrecht 2003, O'Driscoll and Kenworthy 2005). Such situations across the Great Plains further affect people's well-being, and constrain their ability to further cope with other stresses in their socio-ecological system.

Although the challenges can be numerous, tribes have initiated water-related projects that can help them prepare for climate variability and change. These strategies cover a range of actions, which can be identified as assessment, diversification, restoration, and emergency planning. The assessment strategy provides a way to analyze the future needs of a community for various environmental stresses. On the Wind River Reservation in west-central Wyoming, the Bureau of Reclamation examined current municipal and rural water supply systems and wastewater disposal, and also assessed the reservation's future needs (U.S. Bureau of Reclamation 1996). The assessment incorporated water demands for enhanced fire protection capabilities as part of the future needs. Recommendations included the installation of metering to help identify where water leaks in the system were occurring, and the development and implementation of a watershed protection plan to maintain the quality of source waters.

In other communities, actions have been taken to diversify water sources to reduce vulnerability to drought or other catastrophic impacts to their sole water source. On the Rosebud Sioux Reservation in South Dakota, work through the Mni Wiconi-- "Water is life" in the Lakota language-- Rural Water Project (Hall 1998, Rosebud Sioux Tribe 2012) expanded access to the Missouri River sources. Restoration of degraded watersheds and wetlands have also been undertaken to reduce risks to water quality and flood abatement measures. The Potawatomi Reservation in Kansas has worked with Kansas State University to establish several demonstration projects showcasing riparian forest buffers and streambank stabilization techniques for streams that drain cropland. These streams have been subject to erosion and may contain high levels of nutrients and pesticides. Emergency planning has also been effective in reducing risks. In 2007, the Northern Cheyenne Tribe in Montana worked with a consulting firm to develop a Drought Mitigation Plan (Northern Cheyenne Tribe 2007). The plan outlined action items, such as identifying emergency water supplies for each public drinking water system and for the Indian Health Services Clinic. The tribe also plans to continue working with the USGS, EPA, and the Montana Bureau of Mines and Geology to monitor water quantity and quality on the reservation.

HEALTH

Tribes currently face a variety of health care issues, and climate change may act to exacerbate. Expected increases in hot extremes and heat waves may put the elderly and the very young at an increased risk of illness and death (Intertribal Council on Utility Policy n.d., Maynard 1998, Kovats and Hajat 2008). As life spans increase, people in the elderly category will increase as well (Houser et al. 2000). Another group of people vulnerable to heat extremes are those with diabetes (Intertribal Council on Utility Policy n.d., Maynard 1998, Kovats and Hajat 2008). In Native American communities the adult-onset of diabetes has become pandemic (Houser et al. 2000). In tribes in North and South Dakota, one study found the prevalence rate of type-2 diabetes for people aged 45 to 74 to be 33% among men and 40% among women (Lee et al. 1995, Struthers et al. 2003) which is over 4 times the national average. Another factor that makes tribal communities more vulnerable to extreme heat is the high proportion of inadequate housing that provides little protection against excessive temperatures (Houser et al. 2000). Many tribal homes also lack air conditioning or insulation, and residents may not be able to afford the additional costs that air conditioning would entail. Moreover, nationwide, about 14% of Indian households have no access to any electricity, which is ten times the national average (1.4%) (Energy Information Administration 2000).

In addition to extreme heat, other anticipated consequences of climate change in the Great Plains include increases in drought severity and frequency and greater wildfire risks. These factors could lead to a rise in respiratory ailments from increases in dust and smoke (Houser et al. 2000). Asthma sufferers may be particularly vulnerable, and as with diabetes, rates of asthma among Native Americans are higher than the national average. According to the Office of Minority Health, data from 2004-2008 show that American Indian/National Native adults over 18 years of age were 20% more likely to have asthma than non-Hispanic white adults (14.2% vs. 11.6%) and 40% more likely to die (1.3 vs. 0.9 deaths per 100,000).

Climate change health adaptation strategies include programs, such as the diabetes prevention demonstration project of the Winnebago Tribe in Nebraska. This project, sponsored by the Indian Health Service's Division of Diabetes Prevention and Treatment, involves a series of 16 group education sessions using a specially prepared curriculum as well as individual coaching and monitoring (McLaughlin 2010).

Another strategy is a public health campaign, such as the Native American Asthma Radio Campaign, launched by the EPA in 2001 and broadcast in Native American languages, to educate listeners on how to reduce environmental triggers of asthma attacks. Further adaptation measures include the development of tribal energy efficiency codes and weatherization programs (Maynard 1998), the building of new housing units to decrease overcrowding, and the construction of better quality housing units overall to protect against the elements. Improvements in infrastructure, such as road-paving and drainage and strengthening communication links and power supplies, would help decrease health risks from natural disasters (Houser et al. 2000). Recent efforts by Native Great Plains tribal communities include protecting medicinal plants and transporting them to safe areas, developing sustainable agriculture to address nutritional issues in Native diets, obtaining information about social and environmental stress management as climate change action strategies, and obtaining training from the Federal Emergency Management Agency on the development of Emergency Response Plans (Maynard 1998).

ENERGY

Energy concerns on reservations can be framed both in smaller-scale terms of energy use, including supply for residences and vehicles, and in larger-scale terms of energy production as a source of economic development and jobs (see Chapter 6). In a climate change context, energy concerns center primarily around energy usage as a source of greenhouse gas emissions. On Great Plains reservations, many synergies exist for addressing the two sets of small-scale and large-scale concerns.

One of the major concerns surrounding energy usage can be framed in terms of access and efficiency of usage. In many regions, availability of reliable power to many households is lacking. Secondly, due to substandard housing and buildings, energy is wasted in cooling and heating costs. The improvements in affordable and accessible housing materials would greatly alleviate some of the chronic stresses these communities experience.

Development of small- to large-scale energy sources, such as wind, solar, and hydro, would lead to improved access and, possibly, dependable power. This could also lead to improved economic viability of tribal communities if, for instance, tribal wind energy operations were sold through the sale of renewable energy certificates or "green tags" (Gough 2002). Through green tags, the environmental benefits of wind or other renewable energy sources are quantified and sold as a commodity separate from the electricity itself, which is sold as a second commodity with no particular environmental attributes and at a price comparable to its fossil-fuel-based counterparts. An advantage of green tags is that they may be bought by individuals, organizations, or utilities anywhere in the US that would like to support renewable energy development. The tags thus allow consumers to support green power even if their local utility does not directly offer it, and they broaden the potential market for a renewable energy project. The revenue generated through the sale of green tags can significantly boost a project's financial feasibility.

Despite the challenges, the rewards of large-scale tribal renewable energy development in terms of creating long-term sustainable livelihoods, reducing greenhouse gas emissions, and addressing the future energy needs of the Great Plains region could be great both on and off the reservation.

RURAL AND TRIBAL HOUSING

Sustainable, affordable, and energy efficient housing is key for creating community resilience to climate change. It provides major opportunities for both adaptation and mitigation by supplying protection against climate and weather extremes, promoting human health, and reducing greenhouse gas emissions. In the rural Great Plains, there are a variety of housing issues including rural foreclosures, the rehabilitation of housing, the preservation of affordable rental properties, manufactured housing, rural homelessness, and more. Inadequate housing is pervasive among certain groups in the Great Plains, in particular Native Americans and those living along the US-Mexico border.

These communities share certain characteristics, including lower median incomes, higher rates of poverty, and younger populations. According to the 2000 census, the median on-reservation/Oklahoma Tribal Statistical Area (OTSA) Native household income was about $26,700, which was roughly 36% below the national average of $42,000. In border areas, the median household income, as a whole, was $28,000 according to a 2002 Housing Assistance Council report. According to the 2000 census, the percentage of individuals of partial or full Native American descent living below the poverty level on reservations or OTSAs ranged from 13.7% in Kansas to 50.5% in South Dakota, and averaged 26.6% for reservations/OTSAs over the entire Great Plains region. This latter percentage was a little over twice the national average of 12.4%. According to the 2002 Housing Assistance Council report, for the border region as a whole, 18% of residents had incomes below the poverty level. The percentage for Hispanic residents living in non-metro areas was 32%.

Native American Indian reservations are currently suffering from a severe shortage of healthy, safe, and affordable housing, and have been since they were established over a century ago. The need for adequate housing stems back to the 18th and 19th centuries during the eras of removal, reservation and, later, allotment (U.S. Commission on Civil Rights 2003). During this time, tens of millions of acres of tribal lands were either forcibly surrendered or were lost through sales to white settlers. Many native peoples from east of the Mississippi River were relocated from their traditional woodland homelands to unfamiliar, undeveloped, and often barren areas in the Southern Plains. In the

Northern Plains, once nomadic tribes were confined to much smaller portions of their traditional homelands or settled onto lands allotted for farming or ranching, requiring a shift away from tipis to more permanent housing.

In addition to poor building conditions, more than 30% of reservation households nationwide are considered to be crowded and 18% are considered to be severely crowded (U.S. Commission on Civil Rights 2003). Twenty-five to thirty people, for instance, may share a single home (U.S. Commission on Civil Rights 2003). The percentages of overcrowding may be underestimated as no extensive study has ever been done. Also, the census relies on self-reporting, and public housing tenants may not provide an accurate accounting for fear of violating occupancy rules (U.S. Commission on Civil Rights 2003). Homelessness, in which families may live in cars, tents, storage sheds, or abandoned buildings, is also being increasingly observed on reservations (U.S. Commission on Civil Rights 2003). However, no firm statistics for homelessness on reservations are currently available.

In addition to carryover from previous generations, housing continues to be an issue today for a variety of reasons. Many Native communities are geographically isolated and distant from urban centers, which increase the costs of both supplies and labor. Harsh climates may limit the construction season. The construction of public housing on reservations can be very time-consuming because efforts may have to be coordinated among several federal agencies (HUD, BIA, USDA, HHS) and among state agencies as well (U.S. Commission on Civil Rights 2003). Also, there are a variety of complicated and unique land tenure issues in Indian Country. In terms of home ownership, issues, such as predatory lending, insufficient credit ratings, and a general lack of banks and mortgage lenders, are barriers (U.S. Commission on Civil Rights 2003). Additionally, land held in federal trust status, such as land on reservations, cannot be used as collateral for loans. Banks may thus not be inclined to make loans to tribal members for permanent homes, but may provide loans for mobile homes, which they would then have the ability to repossess.

In order to address some of the Indian Country housing issues from the public housing perspective, the Native American Housing Assistance and Self-Determination Act was passed in 1996 separating Native American Housing from other public housing both administratively and financially. The act recognizes Native rights to self-determination and allows the tribes to plan, manage, and monitor housing assistance programs instead of the US government. This should permit each tribe to take into account its unique situation and provide some leeway for tribes to address their housing needs as they see fit. From the private housing perspective, some recommend trying to attract more private mortgage lending to Indian Country (U.S. Department of Housing and Urban Development 1996). However, new housing strategies implemented without simultaneous economic development likely won't work because tribal residents won't be able to pay the rent needed or be able to afford to maintain their homes.

One BIA-funded program on the Crow Reservation in Montana is using the housing shortage as an opportunity to create on-reservation jobs by both producing building materials and constructing high-quality, resilient housing on Crow lands. Awe'-Itche Ashé (Good Earth Lodges) has partnered with the University of Colorado's Mortenson Center to start manufacturing compressed earth blocks using resources from the local area, the location furthest north in the US to do so. Awe'-Itche Ashé is using these blocks to build houses with a passive solar design, thermally efficient windows and doors, and a geothermal system for radiant heating and cooling. The aim is to create long-term, high quality careers for tribal members and create hundreds of sustainable, energy efficient homes on the Crow Reservation.

A second innovative project is taking place on the Pine Ridge Reservation in South Dakota where Oglala Lakota College, the Thunder Valley Development Corporation, the Oyate Omnicye Regional Planning Project, and the University of Colorado's Environmental Design Program are all partnering on a Native American Sustainable Housing Initiative which started in January 2012. The initiative will provide energy efficient housing for Pine Ridge residents and hands-on learning experiences for students. A research component to the project will involve constructing four houses made of different building materials on the Oglala Lakota College campus in Kyle, South Dakota, and monitoring them for indoor air temperature, humidity, and air quality, energy performance, and durability. The homes will be designed with cultural appropriateness as a major consideration, and life-cycle cost analyses will be performed that will account not only for financial costs, but also greenhouse gas emissions, associated with creating the housing materials, constructing the house, and living in and maintaining the house. The ultimate goal of the project is to identify housing options within the community that are healthy, affordable, and sustainable.

Other programs are emerging which provide more financial and technical support for affordable and weather-resilient housing. In the colonias areas, the Nuestra Casa Home Improvement Lending Program of the nonprofit Community Resource Group has been developed to provide better housing. The program is a revolving fund, short-term micro-credit loan system in which a low-income homeowner can borrow $2,500 to be repaid over a two-year period at a 9% interest rate (Giusti 2002, Squires and Korete 2009). The Nuestra Casa program provides a great deal of flexibility in how the borrower can use the funds (Giusti 2002). Another innovative program is Proyecto Azteca's Self-Help New Construction program. Proyecto Azteca is a nonprofit rural housing development organization, based in San Juan, Texas, that serves colonias residents (Arizmendi 2003, Annie E. Casey Foundation 2005a, 2005b). The families receive materials, tools, and instruction, and work together under the supervision of construction trainers to build homes in Proyecto Azteca's construction yard, learning new potential job skills in the process

Human Health and Disease Considerations

POTENTIAL EFFECTS OF CLIMATE CHANGE ON DISEASE: IMPLICATIONS FOR THE GREAT PLAINS REGION OF THE US

In terms of health risks associated with climate change, the primary concern is with infectious diseases (those resulting from the presence and activity of a pathogenic, microbial agent that can be spread among hosts) and *vector-borne diseases* (those resulting from an infection transmitted by blood-feeding arthropods, such as mosquitoes, ticks, and fleas). Only diseases affecting vertebrates will be considered here, but the effects of climate change on plant species can have equally far-reaching effects. In general, these

BOX 8.2

Case Study: Development potential on tribal lands

Tribal residential concerns are often focused around the rising costs of fuel sources used for domestic heating due to poorly insulated housing (Maynard 1998, U.S. Commission on Civil Rights 2003). In order to meet local needs, the Lakota Solar Enterprises founded in 2006, is one of the first 100% Native-owned renewable energy companies in the US and is located on the Pine Ridge Reservation. Lakota Solar Enterprises provides opportunities to reduce both their heating costs and greenhouse gas emissions while simultaneously providing green jobs and training for tribal members, including the manufacture and installation of solar air heaters on Pine Ridge (Koshmrl 2011).

Lakota Solar Enterprises has also been collaborating with the nonprofit Trees, Water, and People to plant wind breaks and shade trees around residences to further reduce energy costs. At the Red Cloud Renewable Energy Center on Pine Ridge, tribal members from all over the US can receive hands-on training in renewable energy applications from Native Lakota Solar Enterprises employees (Koshmrl 2011). In addition, Lakota Solar Enterprises in collaboration with Trees, Water, and People, has implemented the Little Thunder single-home renewable energy demonstration project on the neighboring Rosebud Sioux Reservation, which includes photovoltaic solar panels, a small wind turbine, a solar air heater and a windbreak. These efforts provide additional opportunities in new jobs, more energy efficient housing, and renewable energy sources.

At a larger scale, Great Plains tribal governments and communities as a whole may also be involved in and affected by energy production. Oil and gas operations on tribal lands provide income for the tribal governments in the form of leases and royalties. However, concerns about resulting water pollution and environmental contamination often compete with the desire to develop such resources for the benefit of tribal economic development (Maynard 1998). In some cases, large-scale renewable energy development also has serious impacts on Native communities. Hydroelectric power on the Missouri River has adversely affected Great Plains tribes through the historic relocation of riverside communities, the associated loss of their traditional environs, and the eventual erosion of culturally important gravesites (Gough 2002). Yet, these lands may be ideal for renewable energy production. The Great Plains are home to a phenomenal wind resource on millions of acres of unobstructed, undeveloped land (Garry et al. 2009, Koshmrl 2011). On reservation lands in North and South Dakota alone, the wind power potential is over 240 million BTUs per second (250 gigawatts) (Gough 2002). This is at least one hundred times the hydroelectric power produced by the six large dams on the Missouri River (30). Moreover, development of tribal wind power in the Great Plains could not only reduce greenhouse gas emissions but also help alleviate some of the current and future management demands on the Missouri River (Houser et al. 2000).

However, there are certain considerations in development of energy resources on Native American lands. The Owl Feather War Bonnet wind energy project highlights some of these challenges (Garry et al. 2009). The Owl Feather War Bonnet concerns include consideration of protecting sacred sites and cultural resources (Gough 2002, Garry et al. 2009) in the siting requirements; consideration of tribal council involvement in agreements; and the need for development and access to transmission facilities (Garry et al. 2009). Additional legal issues are associated due to the unique status tribes hold as "domestic dependent nation status" and the access to certain of the existing incentives in further developing energy resource (Garry et al. 2009).

diseases involve a pathogen, one or more hosts, and the environment, which makes these diseases particularly sensitive to changes in conditions. Concerns about infectious and vector-borne diseases in vertebrates can be categorized as affecting:

- Human health because they cause illness and mortality in humans.
- Agricultural health because they cause illness and mortality in livestock and plants, which have direct economic effects on producers and consumers.
- Wildlife conservation and biodiversity because they threaten population viability of native species, especially those that are currently considered threatened and endangered, through changes in life-history traits.

Diseases can be specific to one of these categories or involve all three. For example, West Nile virus is a vector-borne pathogen, introduced into the US in 1993, which causes disease in humans, livestock (primarily horses) and wildlife (primarily birds) (McLean 2008). In addition, wildlife are associated with a number of diseases that are zoonotic (disease normally existing in animals that can infect humans) and play a key role in both the emergence of novel diseases and in the maintenance and spread of pathogens causing currently known diseases. Of the 1,415 infectious organisms known to cause disease in humans, 61% are zoonotic (Taylor et al. 2001, Jones et al. 2008). In addition, the incidence of emerging diseases has increased dramatically since 1940 and, primarily, has been caused by 1) newly evolved strains of pathogens, such as drug-resistant strains of bacteria and the Asian-strain of the H5N1 avian influenza virus; 2) pathogens that have recently entered populations for the first time, such as a corona virus-causing SARS in humans and Nipah virus in domestic swine; and 3) pathogens that have been present historically but have recently increased in incidence, such as Lyme disease in humans (Wolfe et al. 2007, Jones et al. 2008).

Wildlife also plays a critical role in both the emergence and increased prevalence of new pathogens in livestock and humans. Recent increases in incidence of emerging diseases in humans have largely been of zoonotic origin (60.3%), and 71.8% of these were caused by pathogens that originated in wildlife (Jones et al. 2008). In addition, there is an inextricable linkage among pathogens affecting wildlife, domestic animals, and humans, with these pathogens often originating in wildlife and subsequently moving to domestic animal hosts and then humans (Wolfe et al. 2007, Dobson and Foufopoulos 2011). In general, the effects of climate change in creating environments in the US for pathogens emerging outside of the country (e.g., Africa and Asia) have largely been overlooked. For example, if climate change fosters conditions for pathogens, such as Rift Valley fever virus from east Africa (Gerdes 2004), in the US, then introductions of those pathogens are more likely to take hold.

Thus, understanding the effects of climate change on disease requires an understanding of those effects on a wide variety of ecological processes, ranging from pathogen persistence in the environment to vector and host population dynamics to the ability of pathogens to infect new hosts and become established in new environments.

There is general consensus that climate change will affect the geographic distribution of diseases, seasonality of disease incidence, and variation and magnitude of disease outbreaks. However, there is little consensus on how and where this will occur.

While conventional wisdom suggests that climate change will result in the expansion of tropical diseases, especially vector-borne diseases, into more temperate regions (Epstein 2001, Lafferty 2009), there is considerable debate of whether this will occur, at least on a global scale. Randolph (2009) argues that the assumption that climate change will result only in a worsening of worldwide health have become unsubstantiated dogma.

Predictions on the effects of climate change on pathogens and diseases are predicated on the assumption that climate constrains the range of infectious and vector-borne diseases while extreme weather events affect the timing and intensity of outbreaks of those diseases (Epstein 2001). Some of the general hypotheses considered (Harvell et al. 2002) for predicting how climate warming will affect host-pathogen interactions include:

- Increasing pathogen development rates, transmission and number of annual generations;
- Relaxing overwintering restrictions on pathogen life cycles;
- Modifying host susceptibility to infection;
- Disproportionately affecting pathogens with complex life cycles

In general, the effects of climate change are considered to be positive for disease emergence, spread, and incidence. Vector-borne diseases appear to be the strongest candidates for increased abundance and geographic range shifts because many of these are climate-limited with pathogens or parasites that cannot complete development before the vectors die (Harvell et al. 2002). Harvell et al. (2002) also suggest that the greatest impacts of disease due to climate change may result from a small number of emergent pathogens.

CLIMATE CHANGE AND GEOGRAPHIC SHIFTS IN THE DISTRIBUTION OF DISEASES

Vector-borne diseases are especially correlated with changes in climatic conditions (Epstein 2001), primarily in response to the ability of insect vectors to increase in abundance, survive, and transmit pathogens to susceptible organisms. Temperature thresholds generally limit the geographic range of vectors. Expanding tropical conditions can enlarge geographic ranges of vectors and extend the season of pathogen transmission, given precipitation conditions remain equal (Epstein 2001). A number of vector-borne diseases have expanded their geographic ranges into more northern latitudes along with their relevant vectors (see (Harvell et al. 2009)).

Warm nights and warm winters favor insect survival (Epstein 2001), and warm winters tend to facilitate overwintering of both vectors and the pathogens they carry. For example, ticks carrying tick-borne encephalitis and Lyme disease have expanded northward and are predicted to expand even further (Ogden et al. 2008), especially when wild birds are included as a potential transport mechanism for ticks. In addition, conditions during heat waves (high temperatures and high humidity) that often challenge human and livestock health are also the conditions that may favor insect vectors, such as mosquitoes (Epstein 2001).

Of particular concern to human, agricultural, and wildlife health are diseases transmitted by mosquitoes. Dynamic models of the effects of climate change on the global distribution of malaria predicted that climate change will expand the geographic distribution of malaria into North America (Martin and Lefebvre 1995, Martens et al. 1997, Rogers and Randolph 2000). However, the predictions on the extent of this spread vary considerably, depending on model structure and which climate change models were used. For example, Rogers and Randolph (2000) predicted that malaria will occur only in the southern portion of the Great Plains region, whereas Martin and Lefebvre (1995) predicted that, at least under one model, malaria would be more patchily distributed across the entire Great Plains region. Contrary to Epstein (2001) and Lafferty (2009, 2010) argued there is little evidence that existing climate changes have favored infectious diseases. More recent process-based models suggest range expansions or shifts, but little net increase in actual area because increases in habitat suitability for pathogens and vectors have been offset by decreases in habitat suitability elsewhere. This is supported by the models developed by Rogers and Randolph (2000) for malaria spread.

One factor rarely considered in predicting climate change impacts on disease is the effect of restructuring of ecological communities concomitant with changes in environmental conditions that promote pathogen spread and persistence. If climate change reduces the diversity of wild hosts, then pathogens invading a new area will focus on fewer novel hosts and have the capability to have a larger impact, spread further, and have stronger seasonal effects because the 'dilution effect' of multiple potential hosts will be reduced (Schmidt and Ostfeld 2001, Swaddle and Calos 2008, Garrett et al. 2009, Johnson and Thieltges 2010). Thus, there may be synergistic linkages with climate effects on both biodiversity and disease.

CLIMATE CHANGE AND SEASONAL EFFECTS ON DISEASE

In temperate zones, both temperature and precipitation vary seasonally, which has strong effects on disease transmission, especially with vector-borne diseases. Since changes in seasonal patterns are expected with climate change, theoretically this should also affect disease transmission, either in a positive or negative fashion (Lafferty 2009).

There are a number of hypotheses on how climate change could affect seasonal frequency of disease. For example, climate change can lead to an increase in vector abundance while staying within the same seasonal time period, it can extend the season of high abundance, or may lead to a shift in the season of peak abundance to later in the year. Two of the hypotheses were also further explored by Harvell et al. (2002) in terms of R_0 (basic reproductive ratio of a disease), which defines the number of secondary cases produced by an infected individual in an entirely susceptible population. When $R_0 < 1$, the infection will die out in the long run and when $R_0 > 1$, a pathogen will increase and the infection will be able to spread in a population. Hypothetically, increases in temperature not only allow the peak value of R_0 to increase, but also lead to an increased annual duration of the period during which the pathogen is a problem.

CLIMATE CHANGE AND DISEASE OUTBREAKS

While increased warming may encourage changes in geographical distributions of diseases and shifts in seasonal incidence, Epstein (2001) argues that extreme weather events would have the most profound impacts on health issues. However, Pascual and Bouma (2009) point out that variability in infectious disease incidence can be intrinsically cyclic, nonlinear and variable in the absence of any relationship with interannual climate variability. Even so, interannual climatic variability has been shown to influence the size of outbreaks for a number of infectious diseases, especially vector-borne diseases (Pascual and Bouma 2009).

Although higher than average precipitation levels are usually associated with mosquito outbreaks, drought conditions also can play important roles. Landesman et al. (2007) found that West Nile virus outbreaks in humans in the western US were more strongly associated with below-average precipitation in the preceding year. Through wetland surveys and mesocosm experiments, Chase and Knight (2003) found evidence that elimination of mosquito predators in semi-permanent wetlands during droughts allowed populations of mosquitoes to increase substantially in following years, because mosquito predators were unable to recolonize as fast as mosquito production.

Vulnerability, Risk, and Economy; Insurance Industry Perspective

In recent years, the implications of climate change have gained recognition among business leaders worldwide. A prominent example is the insurance and reinsurance sector, which is at considerable risk from the impacts of climate change. These impacts include sea level rise, melting permafrost, floods, heat waves, and an increase in wildfires, drought, and extreme precipitation events (U.S. Climate Change Science Program 2009). Although the scientific community cannot yet prove a definitive link between the planet's warmer climate and individual extreme weather events, the insurance industry has not waited for this causal link to react (Mills 2009).

As the vanguard of risk management, the insurance industry helps society understand and adapt to emerging and evolving risks. Insurers have channeled this expertise into the field of climate change. They have been utilizing data collection, catastrophic modeling, and risk analysis as a means to track trends, define the risks, and formulate solutions for their industry and society at large (Mills 2009). Because of this analysis, they have come to view climate change as a significant cost to their industry, which has resulted in changes in insurance underwriting, investments, and lending credit. A lack of action in response to climate change would constitute a threat to the economy and the insurance industry as a whole (Mills 2009).

The American insurance industry has recently begun to be more engaged in spearheading initiatives and actions on climate change. The National Association of Mutual Insurance Companies have initiated climate change-related action plans and initiatives, and are urging its members to reflect this risk in policies (National Association of Mutual Insurance Companies 2011). Despite the climate-related products and policies now widely available, many insurers initially focused on financial means to limit their exposure to losses related to extreme weather events and natural disasters. This included limiting the availability of policies in certain areas, tightening terms, and raising premiums (Mills et al. 2006).

An example of the industry rationale behind these policy losses and premium hikes can be found with Allstate Insurance, the largest publicly traded insurance company in the United States. Allstate recognizes that there is a relationship between increased extreme weather, catastrophic events, and climate change (Mills 2009). An insurance company that insures one in every nine vehicles and one in every eight houses in the United States, Allstate concedes that climate change contributes to rising temperatures and changing weather patterns. The company believes that these contributions will impact the frequency and severity of extreme weather occurrences and wildfires. Allstate uses this rationale to justify changes in the affordability and availability of homeowners insurance in the US (Mills 2009).

As risks associated with extreme weather events have lowered the availability and the affordability of homeowners insurance in high-risk areas, the responsibility has fallen on the shoulders of the federal government. This scenario is best illustrated by the National Flood Insurance Program (NFIP), which is managed by the Federal Emergency Management Agency (FEMA). The NFIP is a federal subsidy-backed public flood insurance program. It was created in response to a lack of private sector policies for American citizens that live within close proximity to floodplains. Policies are sold by private insurers, but the premiums go directly to FEMA (Drawbaugh 2011). The NFIP has continually been rendered insolvent by extreme weather events.

The NFIP currently is running a deficit of $18 billion and cannot cover its losses without increasing the government's debt burden. In October 2011, the NFIP, which was set to expire in November 2011, was renewed through September 2016. This new bill lowers government subsidies for high-risk property owners, while allowing the insurance industry to raise its premiums in flood areas to reflect the actual risk (Drawbaugh 2011). The insurance industry's heightened participation in the NFIP is expected to strengthen land-use planning and hazard mitigation through market-based signals on risk and remediation (Nutter 2011).

Even though 2010 had a greater number of extreme events than 2011, the total damage in 2011 was more expensive. From extreme drought, heat waves and floods to unprecedented tornado outbreaks, hurricanes, wildfires and winter storms, a record 14 weather and climate disasters in 2011 each caused $1 billion or more in damages and, most regrettably, loss of human lives and property, according to the National Oceanic and Atmospheric Administration (NOAA) (National Oceanic and Atmospheric Administration 2011). The Great Plains experienced damages associated with spring flooding along the Missouri and the Souris rivers in the northern portion, drought and fire losses in the southern region, and tornados in central and southern areas, adding to this total. These occurrences of natural disasters and extreme weather events are consistent with scientific predictions related to climate change.

Thunderstorms, which are common in the Great Plains, are beginning to receive the attention of the insurance industry as high risks. illustrates the increase in frequency of thunderstorms throughout the United States. Allstate is predicting an increase in violent thunderstorms, which are known as "non-model catastrophes" (Lehmann 2011). The insurance company views the increase of these non-model catastrophes as permanent changes and understands the need to recover the costs associated with these events (Lehmann 2011). This permanence will likely be reflected in rate increases for areas affected by thunderstorms.

The insurance and reinsurance industries operate their businesses with the perspective that the climate system is in the process of changing due in large part to human emissions of greenhouse gases. The world's largest insurance and reinsurance companies see the risk posed by climate change as one that poses a risk to their bottom line. While the scientific community is still studying the link between climate change and extreme weather events, these industries have already adapted their business to the realities and uncertainties associated with these impacts.

Insurance and reinsurance companies are sending a clear market signal regarding the economic impacts of climate change. They have changed their risk analyses for extreme weather events and natural disasters to include macroeconomic modeling and catastrophic risk modeling. It is no surprise that, when it comes to reporting on climate change, this industry works hand in hand with the scientific community to develop new risk models for trends deviating from historical realities. Their prioritization of the risks associated with climate change signifies that the insurance and reinsurance industries view the escalating impacts of climate change as definitive aspects of the world's future.

Chapter 9

Collaborative Research and Management Interactions in Response to Climate Change

Since the passage of the U.S Global Change Research Act of 1990, several actions have been carried out in the Great Plains, including development of the first Great Plains regional climate assessment (National Climate Assessment Synthesis Team 2001, Ojima and Lackett 2002), and the establishment of several research centers to support understanding, communication, and response to climate change impacts and consequences. Among these efforts are the Regional Integrated Science and Assessment Centers, National Institute of Global Environmental Change which has been restructured as National Institute on Climate Change Research, North Central Climate Science Center, and other activities supported by state, federal, nongovernmental organizations (NGOs), and local entities.

Actions across the Great Plains have included mitigation efforts to reduce greenhouse gas emissions and sequester more carbon in geologic, soils, and vegetation components of various ecosystems. Recently, managers have implemented adaptation strategies to cope with climate change in local communities, natural resource management, and infrastructure. Given the scope of these activities and the number of federal, state, local and NGO entities involved, there has been little effort made to establish a mechanism for systematic, effective communication, coordination, sharing of knowledge and methods, or co-development of new information to inform decision making, management options, and research directions.

This chapter categorizes activities recently conducted to prepare different sectors and communities for climate change. This is followed by a summary of methods and resources available to further develop responses to climate change. Finally, the chapter presents a framework for greater collaborative and integrative efforts to deal with impacts and consequences, and to develop strategies to meet the opportunities and challenges of climate change, and better monitor and assess the continued climate change impacts in the region.

A variety of factors related to climate variability and change will impact the Great Plains across human and ecological communities. Based on modeled projections of climate change, scientists, land managers, and others are already implementing mitigation and adaptation strategies for agriculture and livestock production and other aspects of the regional and local economy. Response strategies include considerations of ecosystem services that benefit Great Plains communities. In addition, considerations of the impacts on ecological and environmental changes that may affect wildlife and their habitats, including wetlands and river systems are reflected in natural resource adaptation planning. Trends and models also suggest changes in regional climate and the frequency and severity of extreme weather events. Additional impacts include shifts in disease distributions, representing health risks through potential outbreaks. These factors have led the insurance industries to reconsider the elevated economic and human risks and vulnerabilities, complementing scientific research of ongoing and projected climate change.

Response Strategies across a Suite Social-Ecological Dimensions to Multiple Stresses in the Great Plains

Strategies to cope with or adapt to climate change can take multiple forms. Reducing the impacts of hazards and/or anticipatory adaptation before a change occurs is possible as proactive strategies. Coping strategies that Great Plains farmers, ranchers, and other residents may use to deal with climate change include: better preparation for extreme events and multiple-year events, diversification of land-use practices in cropping and livestock systems in order to take advantage of opportunities and reduce vulnerabilities, researching new storage areas for water in case new reservoirs are needed under a changed hydrological regime in the future, and increasing soil organic matter in order to increase water holding capacity and soil fertility. In some areas, Great Plains residents and businesses are also expanding and/or consolidating operations as an adaptation strategy to deal with multiple economic and environmental stressors (Lackett and Galvin 2008).

Climate changes that cause seasonal conditions or extreme events to fall outside of the range of existing coping strategies challenge a system's resilience, adaptability or response capacity. Any system's coping range is spatially and temporally scale-specific, though a goal in vulnerability analysis is to understand where the thresholds might be exceeded to plan for serious consequences of future climate change. Thresholds are characterized by points at which there is a change in the system to cause either increasing vulnerability and/or limited response capacity to some climate disturbance. Events that breach a climatic threshold are thought of as extreme events, although more subtle seasonal shifts can change the conditions where operational strategies and management practices have been developed may be less suitable to these emerging conditions (McNeeley and Shulski 2011). The key to vulnerability assessment is identifying current and potential thresholds for coping with shifts in the average conditions, variability, and extremes of climate. The challenge is capturing the dynamic nature of vulnerability across time and space, and incorporating understanding of future societal changes such as capacity built through adaptations or the damage of cumulative effects and/or multiple stressors.

Adaptive management is one approach that is increasingly modified to cope with and anticipate the impacts of climate change (though few successful examples exist to date). Adaptive management is a potentially useful approach when there is a high degree of uncertainty, risk, and lack of understanding. This approach is especially suited to circumstances where decisions have to be made with a goal of sustaining natural

BOX 9.1
Vulnerability and Risk Conceptual Framing

There is a need now more than ever to develop an integrative approaches to social-ecological studies of vulnerability and adaptation to decision making (Moser 2010). Vulnerability in general refers to susceptibility of social and/or ecological systems to harm from a changing climate, whether through seasonality changes or extreme weather events. Vulnerability to climate change is thought of as a function of a system's exposure, sensitivity and adaptive capacity (Figure 9.1).

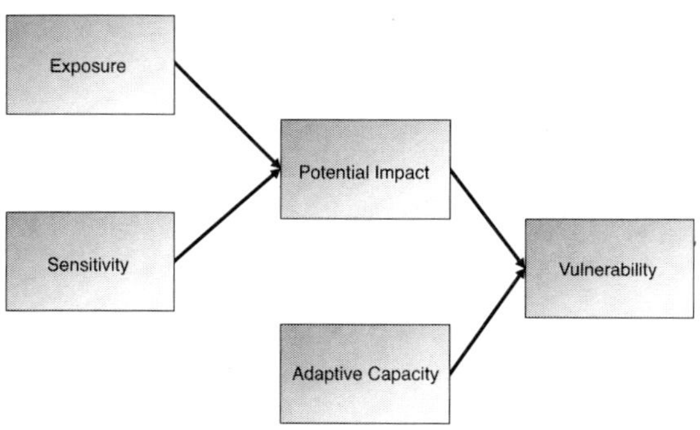

Figure 9.1. Many sectors and disciplines generally perceive vulnerability as a function of exposure, sensitivity, and adaptive capacity, but definitions vary considerably. Adapted from (Glick et al. 2011)

The National Climate Assessment defines these concepts as:

Exposure – in the context of vulnerability to climate change, exposure refers to the climate-related stressors that influence particular systems. This can include stressors, such as drought (e.g., in the context of water resources, agriculture, forestry), sea-level rise (e.g., coastal flooding, habitat loss), or other climate factors;

Sensitivity – defined as "the degree to which a system is modified or affected by (climate) perturbations" (Adger 2006) is a measure of how responsive a particular sector or receptor is to climate variability and change;

Adaptive capacity – is a measure of a sector's ability to reduce impacts through constructive change (Glick et al., 2011).

Vulnerability assessments are important to identifying key vulnerabilities of a region or community or system in order to plan adaptation strategies that sustain livelihoods and ecosystems, and to build resilience to future climate-related shocks. Climate change impacts, vulnerabilities, and risks will differ across various sectors, places, populations, and time scales (Adger, Paavola and Huq 2006, Antle et al. 2004). It is important to identify "determinants" of vulnerability rather than relying solely on "indicators" or "indices", as not all aspects of vulnerability can or should be measured, except in certain cases where places and parameters can be well defined and usually on a local scale (Luers 2005, Hinkel 2011). Determinants of vulnerability are scale-dependent and sector-dependent – i.e., national scale determinants will not be the same as state or local or ecosystem scale (Posey 2009). For example, when looking at indicators for ecosystems at a landscape scale one might use indicators such as landscape diversity and connectivity (Czúcz et al. 2011). In addition, the analysis of vulnerability and adaptive capacity of social systems in a water management context needs to account for variables such as social networks, knowledge of stakeholders, adaptive governance, among others (Huntjens et al. 2012, Downing et al. 2005, Pahl-Wostl 2009).

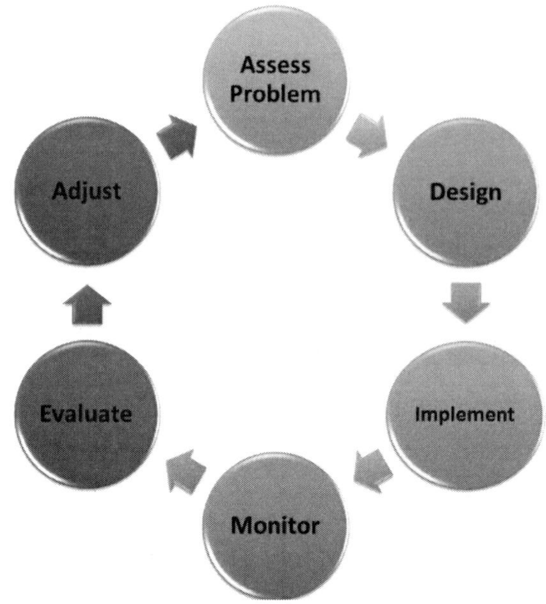

Figure 9.2. Steps for adaptive management. (modified from http://www.doi.gov/initiatives/AdaptiveManagement/)

resources and ecosystem services and their underlying ecological processes (Lal 2001, Folke et al. 2005). An adaptive management approach uses various tools to share and communicate understanding of resource issues among all the stakeholders involved, identifying key uncertainties, exploring alternatives, developing robust policies, and using the outcomes of this process to adapt future policies and actions (See Figure 9.2) (Gunderson et al. 1995). Adaptive management deals with uncertainty by incorporating it as part of the system and using management practices as a tool to gain critical knowledge and experience with dealing with a range of uncertainties associated with the local conditions. Management flexibility is incorporated into the planning process to accommodate various stakeholder interests and to develop strategies that will lead toward "win-win" situations or no-regrets solutions where possible (Johnson 1999). Innovation in planning and implementing management schemes allows for new approaches and ideas to infiltrate these planning and management processes, thereby incorporating learning to guide ongoing management (Pahl-Wostl 2007).

The adaptive management approach first requires that the regional context of the change is understood. Second, "no-regrets" options -- ones that make sense given current conditions and potential future ones -- should be identified and considered through the use of scenarios. Third, people need to be provided with practical and tractable alternatives for adaptation. Fourth, decision makers should "learn by doing" and evaluate results along the way, making the process of adaptation an iterative process. And, last, the public must be kept informed of the implications of change. The approach also requires the creation or support of the appropriate institutions and collaborative learning mechanisms that include local stakeholders, managers, researchers, policymakers and other stakeholders that can help satisfy multiple goals, such as achieving conservation goals while producing community benefits (Bosch et al. 2003, Berkes 2004). Top-down management approaches which also rely on prescriptive strategies are poorly suited to meet complex multi-sectoral problems (Berkes 2004). Adaptive management approaches calls for cross-jurisdictional considerations between local to regional agencies to promote communication, knowledge sharing, and learning within and between various stakeholder organizations (Berkes 2004).

The Department of Interior provides technical guidance documents for land and natural resource managers on implementing adaptive management approaches.

ASSESSMENT OF CLIMATE CHANGE RESPONSE STRATEGIES

Research efforts have brought attention to the role ecosystems have in providing key economic goods and the ecosystem services that sustain, regulate, and support life on Earth (Costanza et al. 1997, Daily 1997, Daily and Ellison 2002). Terrestrial ecosystems provide a wide array of goods and services that human well-being, and even survival, depends upon. Consumptive goods provided by land systems, such as grains, animal protein, and fiber and wood products, are typically valued through well-developed markets. But the societal and ecological contributions of the 'underpinning' services provided by ecosystems often remain 'invisible' and unvalued (or undervalued). The array of such services is broad, from those services that regulate critical human-environment processes (e.g., climate, disease, flooding, detoxification) to services that support economic activity (e.g., soil formation, primary productivity, nutrient cycling, pest control, pollination).

Incorporating ecosystem services into the decision making process allow managers to better understand the effects of land use and management. In addition, evaluating changes in the state of ecosystem services (i.e., soil fertility, water resources, and food and fiber production) is critical to the development of appropriate coping or adaptive strategies under different human-environmental stresses. Some impacts may present only temporary disruptions. But in some cases; such as the plowing out of grasslands, conversion of land reserves, draining of wetlands, or introduction of novel species for bioenergy production; the impact on ecosystems can be more long-term and affect the integrity of these systems to a point at which a transition to a less desirable or less productive state could occur. Recognizing the importance of change as a basic component in managing for climate change is essential for developing resilient and more robust adaptation strategies (Berkes and Folke 1998, Gunderson and Holling 2002, Walker and Meyers 2004, Tschakert et al. 2007).

Available case studies provide a wealth of data on the social, biological, and physical components of coupled human-environment systems. Data from intensive case studies can enable evaluation of conditions determining the vulnerability or resilience of systems to different scenarios of social and environmental conditions. Characterizing and determining ecological thresholds are challenges to resource managers and to society, due to the sudden and often irreversible nature of the changes in ecosystem services and the new conditions that emerge (Hobbs et al. 2006). Socio-ecological thresholds due to interacting environmental and socio-economic drivers are being triggered in many semi-arid systems around the globe. Basic understanding of where and when a threshold will be crossed is still unclear (Julius et al. 2008), however, the inherent sensitivity of semi-arid systems to climate variability and land use changes has been documented (Ojima et al. 1993, Parton et al. 1994, Archer et al. 2001, Reynolds et al. 2001). The fragmentation of landscapes and the discontinuity of landscape processes also contribute to ecosystem and biodiversity vulnerability in ways that contribute to triggering social-ecological thresholds (Lackett and Galvin 2008).

FORECASTING AND OBSERVATION TECHNOLOGY CONSIDERATIONS

Forecasting technologies have advanced; linking field observations, remote sensing, and modeling systems; to provide seasonal forecasts of crop and ecosystem productivity. Monitoring of key climate characteristics (e.g., temperature maxima or minima, seasonality of precipitation patterns, and interactions among climate characteristics) combined with improved productivity forecasting capabilities provide greater forewarning of impending critical thresholds in response to extreme events in currently functioning landscapes and ecosystems. In addition, observations of ecosystem indicators associated with biotic assemblages (e.g., host-pest relationships), ecosystem functions (e.g., water use efficiency, nutrient cycling, and carbon assimilation), and structural changes (e.g., woody to herbaceous ratio, bare soil exposure) can provide clues to emerging thresholds. A number of ecosystem services can also be monitored to assess the impacts of change to society and vice versa.

Improved understanding of ecological thresholds is gained through an integrative, prognostic approach which integrates a suite of observations as a basis for forecasts of the probability a threshold is being crossed (Walker and Meyers 2004, Hobbs et al. 2006, Lyytimäki and Hildén 2007). Social-ecological vulnerabilities can be assessed using various approaches (Moss et al. 2000, Turner et al. 2003, Adger et al. 2004, 2007, Fussel and Klein 2006, Smit and Wandel 2006, Ford et al. 2010)(Moss et al. 2000, Turner et al. 2003, Adger et al. 2004, 2007, Fussel and Klein 2006, Smit and Wandel 2006, Ford et al. 2010). These approaches provide a framework to correlate social outcomes, such as poverty reduction, against measures of capital assets or other measure of available resources (Eriksen and O'Brien 2007, Eriksen et al. 2007). Other approaches analyze societal needs (e.g., food availability, water access, health care) in the context of various stresses, such as commodity price volatility or climate variability (Luers 2005, Thomas and Twyman 2005, Lal et al. 2011). Such approaches allow for better integration of environmental and societal metrics and variables to evaluate social-ecological vulnerability. The choice of coping strategies is determined by the capital resource assets (i.e., natural, human, social, financial, and built capital) available to different community members in a particular location and time (International Institute for Sustainable Development et al. 2003). Decisions are based on multiple criteria related to various cultural worldviews, economic and other values, and societal goals for various communities (Etkin and Ho 2007). Regional and local decisions to cope with stress and to overcome vulnerable conditions will reduce the impacts of these stresses and make decisions that will benefit some and affect others differentially (Adger et al. 2005, Dolan and Walker 2006).

ADAPTATION AND MITIGATION STRATEGIES

Climate change adaptation in human societies requires responding to climate stimuli, such as recent events in the Great Plains associated with extreme weather events, floods, and droughts, but -- perhaps more importantly at this point in time -- also anticipating and planning for potential changes (Smit et al. 2000), especially where early warning signs are present (Glantz 1999). Adaptation here refers to a fundamental, systemic change in response to environmental conditions, that maintains or strengthens the viability of the system (Smithers and Smit 1997). Climate change adaptation of social-ecological systems needs to operate across local to global scales, and requires the proper functioning of social, ecological and institutional systems. Thus, sustainable adaptation emphasizes strategic, collective action in response to or anticipation of harmful climate change to minimize disruption to key resource flows and adverse effects on human and ecosystems well-being. In other words, adaptation enhances the ability of the natural environment to meet current needs and also continue to provide ecosystem services for future generations (O'Brien et al. 2004, Eriksen and O'Brien 2007, Eriksen 2011, Eriksen and Brown 2011, McNeeley 2011).

Adaptive capacity is constrained by factors that restrict people's set of options when social-environmental conditions change (Berkes and Folke 1998, Gunderson and Holling 2002). Institutional hierarchies may constrain adaptive mechanisms operating at the community level, due to policy directives developed for conditions in which climate change considerations were not accounted for (Adger and Kelly 1999). Institutional responses to climate change are often best suited for mitigation of emergency situations and isolated events, rather than for slower onset, cumulative or systemic climate-related problems leading to disruption of ecosystem services. Institutional and regulatory entities are even less well-suited to working with underlying social factors that determine vulnerability (Handmer et al. 1999). Where institutional rule-making occurs in a compartmentalized and fragmented framework, responses to climate change have been either nonexistent in the worst case, or case-based mitigation in the best case (McNeeley 2011).

Response to environmental vulnerability and risks is typically determined by a series of livelihood decisions that depend on the community or household assets and the allocation of these assets to generate benefits and well-being for various stakeholder groups (Kelly and Adger 2000, Barrett et al. 2001). Adaptation actions are choices within a "response space" that includes coping, but also longer-term adaptation actions. In these situations, successful actions promote system resilience, promote legitimate institutional change, and hence generate and sustain actions (Osbahr et al. 2010). Decisions, in reality, are constrained by the broader economy and political milieu, as well as by prior decisions that partly lock people into particular livelihood pathways. Actions are driven by objectives, such as income diversification, risk minimization, and capital accumulation (Allison and Hobbs 2004, Lorenz 2010) and informed by their perceptions of how the natural world, including climate, functions over time (Douglas and Wildavskky 1982, Thompson et al. 1990, Verweij et al. 2006). Adger et al. (2009) assert that adaptation has social limitations, yet does not have to be limited by uncertainty of future risks (Adger et al. 2009). In that case, what are the opportunities for adapting natural resource management and livelihood strategies for climate change in the Great Plains region?

Approaches to Enhance Great Plains Climate Change Research and Adaptation

Land use and other resource decision-making processes provide a foundation for evaluating factors that influence human activities and their effects on ecosystem services. The relationships of the coupled human-environment system can be defined through the nexus of the decision-making process and delivery of ecosystem services. The environmental context of the system can be determined by the state of ecosystem services and the reliance of the decision maker on these services. Instability in the system may arise when unforeseen loss of an ecosystem service occurs, such as loss of soil stability and vegetative cover during a drought resulting in a massive dust storm, as in the 1930s Dust Bowl, or lack of water flow leading to desertification or diminishing stream flow, for example, the Rio Grande River not always reaching the Gulf of Mexico. The effect on the coupled human-environment system may seem to appear rapidly, although the underlying changes have been occurring over time (i.e., "creeping environmental problems"), undetected until a critical threshold had been met (Glantz 1999, Smit and Wandel 2006).

From the rich literature on developing and implementing climate adaptation, we can identify a small set of common principles (Willows and Connell 2003, Hansen and Hoffman 2010, Halofsky et al. 2011, Mawdsley 2011). A first principle is that the scope and scale of climate impacts and adaptation typically require considering issues expressed at multiple scales of space, time, and complexity. These issues must be addressed by decisions that occur in very different ecological, economic, social, and organizational contexts. To do so, it is usually necessary to involve communities and decision-makers at multiple scales appropriate to addressing changes in the social-ecological system. A diverse community of participants facilitates identification of the full range of issues and potential policy and management decisions (Joyce et al. 2009, Adger et al. 2011, Robinson and Berkes 2011). This integrative approach incorporates uncertainty and risk assessments, links modeling analyses and decision making at appropriate spatial and temporal scales, and provides a mechanism for sharing resources and knowledge across affected communities and planners (Joyce et al. 2009, Ojima and Corell 2009). It explicitly recognizes that climate change impacts cross jurisdictional and disciplinary boundaries, and that effective partnerships are essential for addressing climate change.

Common traits of climate adaptation planning processes are illustrated in Figure 9.3 (adapted from National Park Service 2010). This climate adaptation framework incorporates elements common to more traditional adaptive management (e.g., Holling 1978; Williams et al. 2007), and is useful to identify an integrated set of activities that lead to effective climate adaptation. While this framework presents these activities in a logical order, specific activities will occur when the opportunity presents itself in most cases, rather than in the linear order suggested by Figure 9.3. This framework articulates key steps that apply generally to decisions under high uncertainty, and specifically to decisions under rapid climate change.

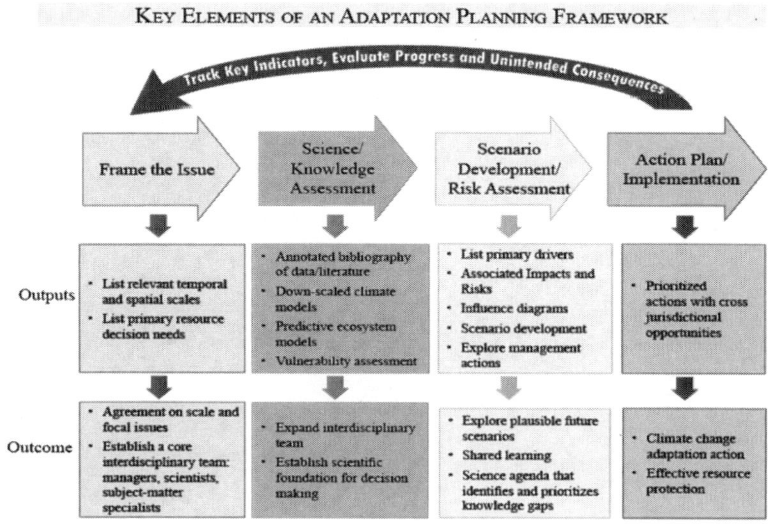

Figure 9.3. Conceptual framework for collaborative adaptation planning (modified from National Park Service (2010).

The first steps -- noted under "Frame the Issue" -- focus on identifying specific concerns and issues the community faces within the prescribed social, ecological, and scale-dependent context (Smit and Wandel 2006, Ojima and Corell 2009). Appropriate scales of analysis in space and time can be identified with assistance from the stakeholder community and by incorporating local knowledge and observations. The time scale of analysis needs to be matched to issues that respond at different scales, such as forage growth and livestock production, filling of reservoirs and other hydrological responses, vegetation recovery or transition, or other climate triggers that affect maintenance of infrastructure or delivery of ecosystem services. This phase of the adaptation planning process emphasizes key resources and values.

Assessments are logically conducted once the scale and scope of the issues have been articulated (Figure 9.3). An assessment can focus on a specific geographical unit, sector, or domain defined by key resources. The Badlands National Park vulnerability assessment includes the park and surrounding landscapes necessary to support biodiversity and other processes (Hansen et al. 2011). This assessment is unusual in that it includes species, habitats, selected infrastructure, and cultural resources. Other Great Plains assessments have addressed key species (Zack et al. 2010), crops (Weiss et al. 2003), fish (Perkin and Gido 2011), water (Stone et al. 2003), or other sectors. River basins often define relevant social, economic, and ecological units. Northern Great Plains river basins are the focus of an ongoing assessment that involves climate model downscaling, runoff modeling, and an assessment of ecological consequences (Skagen and Melcher 2011).

Vulnerability assessments and other activities in the second column of Figure 9.3 focus on synthesizing and evaluating information that helps identify resources at risk, why they are at risk, and the information that guides these evaluations. While these processes inform decisions, other steps and activities are necessary to identify potential management or policy actions, and select or rank alternative decisions and actions.

Activities in the third column of Figure 9.3 focus on identifying and ranking potential actions and decisions. A variety of methods can be used in this process, and scenario development, as identified in Figure 9.3, is only one of the alternatives (Peterson et al. 2003, Willows and Connell 2003, Williams 2007, Nichols et al. 2011). While existing processes will contribute to risk assessments and decisions on climate adaptation, considerable work is needed in this area to address the different ways of knowing and understanding risk and uncertainty (Eakin and Patt 2011, Pidgeon and Fischhoff 2011, Renn et al. 2011).

The final step is to design and implement adaptation plans that address changes in climate and effects on the socio-ecological system. In almost all cases, the planning strategy will be a recursive or iterative process (Dessai et al. 2005, Fussel 2007, Wilby et al. 2009, Preston et al. 2011). Any or all steps in the planning process may need to be revisited as information accumulates and priorities change (Jones and Preston 2011).

Threshold changes in the socio-environmental system are difficult to predict, and policy- or decision makers may not always be able to anticipate all impacts (Lyytimäki and Hildén 2007). Effectiveness of selected actions is highly dependent on institutional setting and level of engagement by institutions in the planning and implementation processes (Lyytimäki and Hildén 2007). Development of strategies and analysis of thresholds are only effective if there is an appropriate set of agents or institutions to take appropriate actions.

IMPACT STUDIES AND CLIMATE ANALYSIS (MONITORING OF SOCIAL-ECOLOGICAL SYSTEMS, SYNTHESIS STUDIES, THRESHOLD ANALYSIS)

Effectively addressing climate change and its effects on ecosystems, resources, and society will require coordination in the research and observation capabilities of multiple organizations, institutions, and government programs. Many organizations have ongoing monitoring and evaluation programs relevant to detecting and responding to climate-driven changes (Table 9.1. Monitoring). A key issue is that each agency has developed monitoring systems with a specific mission orientation, but these systems have not always had climate change effects in mind. Now that efforts are underway to organize observations to specifically address climate change, it is critical to evaluate the manner in which monitoring systems provide information and to seek synergies among the various monitoring efforts to develop a comprehensive system of observations and assessments. Figure 9.4 illustrates a functional, integrated system, consisting of multiple observing systems, modeling, and evaluation components that address the societal need to assess sources and fluxes in CO_2 and other carbon pools.

An important feature of the ecosystem carbon models is that they can be compared to the time-varying concentrations of atmospheric carbon dioxide as a means of model validation (Carbon Tracker). This comparison requires bridging across model response time scales, where measurable atmospheric variations occur on much shorter time scales (seconds to days) as opposed to measurable ecosystem fluxes and stock changes (hours to decades). Even with this timescale mismatch, the atmospheric constraints provide an important test for ecosystem model predictions.

Table 9.1 Examples of existing Federal programs that monitor and evaluate Great Plains resources and processes relevant to assessment of climate changes and vulnerabilities.

Organization & Program	Relevant Foci	Reference
U.S. Department of Agriculture		
SNoTel	Snow and water monitoring.	http://www.wcc.nrcs.usda.gov/snow/
Animal and Plant Health Inspection Service (APHIS)	Pests, diseases surveys and monitoring.	http://www.aphis.usda.gov/
Natural Resources Inventory (NRI)	Land use, land cover, erosion, wetlands.	http://www.nrcs.usda.gov/wps/portal/nrcs/main/national/technical/nra/nri
Forest Inventory and Analysis (FIA)	Forest extent, composition, condition, invasive species	http://www.fia.fs.fed.us/
Agricultural Statistics Service	Crops, demographics, livestock, economics	http://www.nass.usda.gov/index.asp
U.S. Department of the Interior		
National Water-Quality Assessment Program (NAWQA)	Monitoring and assessment of ground and surface water composition, attributes, and quality.	http://water.usgs.gov/nawqa/
National Water Information System	Real-time and historical flows, levels, meteorological data, and associated attributes of surface and subsurface waters	http://waterdata.usgs.gov/nwis
Inventory and Monitoring Program	Key indicators of natural resources in national park units	http://science.nature.nps.gov/im/
Environmental Protection Agency		
Climate change indicators	Greenhouse gases, climate indicators, ecosystem responses	http://www.epa.gov/climatechange/indicators.html
National Oceanic and Atmospheric Administration		
National Weather Service	Local to global weather, hydrology, storms, and other hazards	http://www.nws.noaa.gov/
National Integrated Drought Information System	Information on historical, current, and emerging drought	http://www.drought.gov
MultiAgency		
National Land Cover Data	National land cover trends (MRLC)	http://www.mrlc.gov/
National Atmospheric Deposition Program (NADP)	Atmospheric composition, . deposition	http://nadp.sws.uiuc.edu/

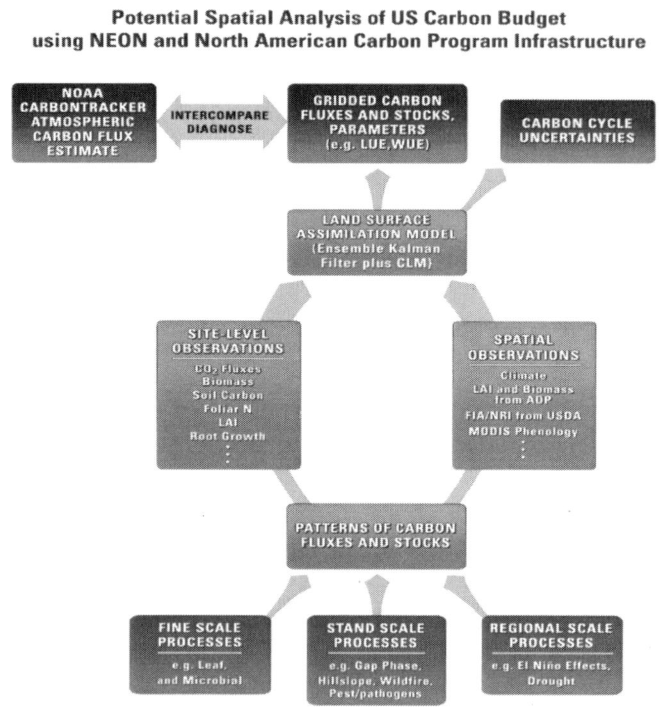

Figure 9.4. A conceptual analysis and forecast of the US ecosystem carbon budget derived from multiscale observations and an integrated carbon assimilation model. LUE = Light use efficiency, WUE = water use efficiency, CLM = the NCAR Community Land Model (Bonan et al. 2002), LAI = leaf area index, FIA = Forest Inventory and Analysis of the USDA, NRI = Natural Resources Inventory of the USDA, MODIS = the Moderate Resolution Imaging Spectroradiometer satellite instrument, Foliar N = foliar nitrogen. CarbonTracker is a NOAA tool that estimates carbon fluxes from atmospheric CO2 measurements and related meteorology (Schimel et al. 2009).

NATIONAL PARK SERVICE CLIMATE ADAPTATION ACTIVITIES AND NEEDS WORKSHOP

The National Park Service recognizes that gaps in climate literacy of staff and stakeholders are a significant limitation to identifying and implementing climate adaptation actions. Workshops with presentations by local, regional and national managers and scientists have been held to provide general information, facilitate relationships within and between organizations, and help motivate actions focused on specific locations. Two recent examples from areas with very different resources and circumstances clearly illustrate the need for these workshops, and serve to identify ongoing activities that will likely be necessary to make more rapid and efficient progress toward climate adaptation.

Case 1: Rocky Mountain National Park and the surrounding Arapahoe, Roosevelt, and Routt National Forests are located in northern Colorado, close to more than 1 million people, four major research universities, and a plethora of research agencies. These areas receive intense recreational use and have been the focus of many short- and long-term studies. The region also has a wealth of local scientific expertise and knowledge. In addition, considerable effort has been directed to establish and maintain working relationships between the federal, state, county, and local municipalities and organizations. Drawing on this reservoir of talent and infrastructure, a climate adaptation workshop held in November 2010 presented information on climate adaptation, facilitated interactions among participants, and identified and documented priorities and opportunities for better multi-agency coordination and collaboration (Thompson, 2010). Workshop

BOX 9.2

Scenario Applications: Using Scenarios to Explore Assumptions and Test Management Alternatives as Conditions Change

Scenarios are plausible, internally consistent stories about the future, challenging us to consider how we would operate under novel conditions. Scenario thinking is a structured process by which groups can organize perceptions, assumptions, and complex data about how the future may evolve over time into sets of scenarios. Managers can then use the information to explore unknowns, test strategies, generate new ideas, improve organizational flexibility, or inform decision making in situations of risk, uncontrollability, complexity, and uncertainty.

The US National Park Service and partners are using multivariate climate change impact scenarios to address future risk and uncertainty in resource management. The National Park Service develops scenarios through a participatory process that integrates quantitative, model-driven climate change data with qualitative and practical information about how environmental impacts and future socioeconomic conditions could interact and affect park resources and operations. The resulting multivariate scenarios allow resource managers to explore and understand the range of potential future environmental, social, and economic conditions, and to develop flexible management actions and strategies in spite of uncontrollable and irreducible uncertainties.

In its initial application through the National Park Service, the agency's scenario development and application has proven successful at fostering rich interactions between climate scientists and decision makers, broadening decision makers' perceptions of potential climate impacts, inspiring robust management actions and strategies, and identifying inefficient or counterproductive management policies and actions.

The National Park Service is continuing to develop and refine methods for applying scenarios to management questions, but the scenario planning techniques developed to date are already being incorporated into the agency planning framework, and are helping to evolve that framework to support adaptive management. Moreover, the National Park Service staffs are using the compelling place-based narratives generated during the scenario process to communicate climate change information with a variety of audiences, from National Park Service scientists and facility managers to park visitors, stakeholders, and the general public.

Case Example: Wind Cave National Park

In 2009, the National Park Service conducted a scenario-thinking project that focused on Wind Cave National Park in South Dakota. Researchers and resource managers used downscaled regional climate projections, published information on potential climate change impacts in the Midwest, and national socioeconomic trends to develop a set of four, park-scale, multivariate climate change impact scenarios. Park managers and researchers used these scenarios to identify threats to park resources and operations, areas for additional research, and opportunities to foster resiliency in park resources, operations, and infrastructure. Specifically, managers discussed potential threats to park resources and operations, such as water resource shortages and archeological resource exposures; areas for additional research, such as climate change effects on cave environments or climate-induced changes to visitation; and opportunities for strategic capacity building, such as integrated research and monitoring partnerships with local universities, agencies, and volunteer groups.

BOX 9.3

Case Study: Scientists and Managers Working Together to Find Solutions for our National Grasslands

Recent efforts to engage a more effective dialogue between resource managers and researchers have taken place around the country, including the Great Plains. One such effort was recently conducted by the US Forest Service, through coordinated efforts of the Rocky Mountain Research Station and the Rocky Mountain Region of the Forest Service. The program included researchers from the Agricultural Research Service, Forest Service, and academic institutions. This effort initiated knowledge sharing among Great Plains scientists and managers on the topic of climate change by hosting two events: a day-long webinar on science findings and a follow-up workshop via video teleconference with a group of invited resource managers and scientists to discuss critical issues identified by managers.

Webinar and Workshop

The webinar engaged Great Plains managers and scientists with a common interest in the future of the region's grasslands. Experts on climate change effects applicable to Great Plains grasslands presented research findings. Participants represented a broad range of affiliations, from federal and state agencies to nonprofit organizations, universities, and private consulting firms. Recordings of the presentations are available at http://www.fs.fed.us/rm/grassland-shrubland-desert/events/climate-change-webinar.

The goal of the follow-up workshop was to identify products or tools needed by managers to promote sustainability of national grasslands in the face of climate variability and change. National Grassland managers were particularly interested in presentations that gave specific guidance or suggestions for management (e.g, types of vulnerability assessments, technology applications) or that predicted outcomes of complex interactions (e.g, plague and prairie dogs, vegetation shifts, demographics). A number of participants highlighted talks on vulnerability and risk assessments, including as integrated assessments that consider multiple sectors simultaneously. Participants noted topics of interest that were not covered. One was the inclusion of a more global perspective on climate change, including how global change may affect local issues and vice versa, and what lessons could be learned from international efforts. Another requested topic was the addition of more detailed water projections, including changes to aquifers and effects on aquatic species. Along with needs and priorities, workshop participants shared barriers to effectively integrating climate change into decisions. These included:

- Uncertain and limited funding
- Lack of knowledge on how to manage grasslands for resilience
- Lack of guidance on how to apply projected climate change effects to management decisions
- Inertia and resistance to shifting from old management strategies
- Politics distracting from integration of climate change into programs
- Different land ownerships and policies on adjacent lands
- Lack of knowledge on how to accommodate variability, extreme events, and uncertainty in management decisions
- Large number of existing stressors in a highly fragmented landscape with many species in decline

Solutions

Both scientists and managers suggested solutions or products that could reduce management barriers and improve climate change response. The

BOX 9.3

Case Study: Scientists and Managers Working Together to Find Solutions for our National Grasslands (cont.)

reinforcement of partnerships was a common theme that was promoted through the workshop. Managers noted the need for a centralized mechanism to communicate current and ongoing research projects in the region. Products that promote education and awareness of local climate change issues, including additional webinars and workshops, were seen as critical to engage stakeholders and inspire action. Climate change can also present opportunities; for example, carbon sequestration can be a driver for implementing grassland restoration projects. Other participants suggested more specific measures, such as implementing changes in breeds or species of grazers to cope with changes in forage productivity or composition. New technologies can be integrated into management, such as the transmission of real-time remote-sensing data through wireless devices to better inform day-to-day management decisions, or use of social networking to bring together stakeholders.

Lessons for Scientists

Publishing research findings in scientific journals and presenting at conferences primarily attended by colleagues will not adequately disseminate information to managers. Consultation with managers during the study design phase can improve the utility of research findings to on-the-ground actions. Finally, National Grasslands provide opportunities for climate change studies.

Lessons for Managers

Although preparing for climate change may seem daunting, managers can start with current management strategies that are applicable to climate change issues, such as reducing potential for soil erosion and protecting riparian corridors. Scientists are eager to help managers and to see their research applied, but are often hesitant to extrapolate findings to make specific management recommendations. Managers need to discuss options and include input from scientists during planning phases. Managers also need to be aware of the limitations of individual studies or assessments and how that affects their applicability to local issues.

Next Steps

The workshop and webinar served as a catalyst for creating a productive partnership that uses a science-based approach to incorporating climate change into land management. Finding climate change solutions and encouraging dialogue among scientists, managers, and stakeholders requires an ongoing effort. Having created momentum through the workshop, the core group of participants must plan for efforts to continue the engagement and to update each other on research proposals, science findings, and products relevant to the Great Plains grasslands.

participants generally knew (or knew about) each other, but an additional effort would be required to organize and coordinate the multi-agency and multi-disciplinary groups and activities that would best address broad-scale climate adaptation needs. The 2010 workshop was very well attended, received high praise, and generated excitement and enthusiasm. However, identifiable follow-on events failed to materialize during the following year. Participants would surely agree that the workshop increased climate literacy and knowledge, but the follow-on activities needed to sustain momentum and enthusiasm have not occurred.

Case 2: Black Hills and surrounding parks, forests, and grasslands located in western South Dakota, including Wind Cave National Park and Badlands National Park. More than 90 participants from the Great Plains attended a workshop in April 2011 in Rapid City, South Dakota (Thompson 2010). The workshop included presentations and group activities that facilitated learning, and identified climate related priorities and follow-on actions. In comparison to the region surrounding Rocky Mountain National Park, Rapid City is isolated, close to a small population, has access to few local climate experts, and is the focus of a far less intense research effort. But a year after the workshop, the long-term results have proven similar. Participants clearly gained a better understanding of climate issues, but further engagement is needed to sustain action.

These results emphasize a key issue: land management staff, generally, does not have the time or resources to organize and sustain the broad-scale, multi-agency, multi-disciplinary efforts that best facilitate climate adaptation. There is a key need for help to establish and sustain community-level activities. The goals of the community-level activities should include empowering local residents and organizations and providing specific expertise where needed. Managers need help to organize meetings, facilitate activities, and expand the geographical and disciplinary scope of work that is necessary to identify, implement, and sustain climate adaptation. The National Park Service and other organizations are conducting activities focused on parks or other units, and these activities usefully contribute to broader-scale efforts. But staff assigned to specific units do not have adequate resources -- and often not the authority -- to organize and motivate broader multi-stakeholder communities.

Case 3. Eastern New Mexico Carbon Sequestration study organized by a small group of northeastern New Mexico ranchers, working in collaboration with the National Carbon Offset Coalition. Several ranchers banded together to apply to the Chicago Climate Exchange for a rangeland carbon offset project. Chicago Climate Exchange has published protocols for the organization, implementation and verification of rangeland carbon offset projects. The ranchers, with technical support from USDA-NRCS, ARS-Jornada Experimental Range, the Department of Energy Southwest Regional Carbon Sequestration Project and New Mexico State University provided baseline information, 5-year management plans and monitoring schemes to meet the protocol requirements. The focus of the project was proper livestock grazing management, including factors such as stocking rate, distribution and season of use. The goals were to maintain net primary productivity within the herbaceous component of plant communities, provide adequate fine fuel to allow for strategic burning to reduce shrub cover, and minimize losses of soil carbon during drought periods. Although the prices for greenhouse gas mitigation activities have been at historic lows over the life of the project, the ranchers involved have received annual payments (see De Steiguer et al. 2008 for a description of the project).

REGIONAL EXAMPLES OF ADAPTATION PLANNING ACTIVITIES ACROSS THE GREAT PLAINS

In the agricultural sector, farmers are experimenting with various conservation strategies to adapt and cope with climate variability and change. Some of these include conservation tillage systems and methods to retain soil organic matter to limit erosion and increase water retention capacity (Knutson 2008). Another transition many farmers are making is switching from flood irrigation to sprinkler or drip irrigation systems to conserve water. However, in some surface-irrigated basins this has the unintended consequence of reducing return flows that are important to both the riparian ecosystem as well as downstream users. As part of the Environmental Quality Incentives Program in Nebraska, the USDA National Resources Conservation Service (NRCS) has a special initiative in the water-stressed Pumpkin Creek watershed where farmers are offered financial incentives to transition from irrigated to dryland cropping (Knutson 2008). This program has helped reduce groundwater pumping to restore the watershed, and it is helping the region's residents to proactively adapt to a drier future rather than having to cope with a transition toward rainfed agriculture under crisis conditions (Pope 2007). Livestock producers in the Great Plains are also experimenting with new strategies, such as rotational grazing where cattle are rotated to smaller pastures to allow for grass regeneration (Knutson 2008).

The NRCS administers a variety of conservation cost-share and technical assistance programs that could be refined and redirected to more effectively cope with climate change. From 2005 to 2009, NRCS rangeland-based conservation programs provided almost $130 million to private landowners to improve management in 6 central US states (Kansas, Nebraska, North Dakota, Oklahoma, South Dakota and Texas), which compose the majority of the Great Plains (Tanaka et al. 2011). The expenditure supported application of defined conservation practices, such as brush management, prescribed grazing, range planting, riparian buffers and wildlife habitat management, affected management on more than 30 million acres (12 million hectares).

In addition to the cost-share funds provided, significant amounts of technical assistance were provided to support the proper application of mechanical and management technologies. Although there were likely significant benefits derived from the application of these practices (Briske et al. 2011), climate change mitigation and/or adaptation were not explicitly considered in program design. Similarly, there is a relatively poor quantification of the impacts of existing conservation programs on climate change response (Bestelmeyer et al. 2011). Including consideration of climate change projections would enhance the robustness of these or similar programs. In addition, enhancing communication between ongoing programs and research across the Great Plains could greatly improve the cost-effectiveness of public expenditures for conservation and adapting to climate change (Briske et al. 2011).

Denver Water. On the western edge of the Great Plains, Denver Water has implemented a host of innovative strategies for drought planning, climate change adaptation, and conservation strategies (Denver Water 2012). This includes an Integrated Resource Plan, initiated in 2008, to guide efforts for the next 40 years. In addition, the utility is negotiating a historic collaborative water-sharing agreement -"Colorado River Cooperative Agreement: Path to a Secure Water Future" - with a number of Colorado West Slope entities to ensure sustainable water resources in the uncertain future. This visionary agreement proposes three main areas to move forward from conflict to adaptive collaboration: 1) resolution of historic conflicts and a holistic approach to resolving Colorado water disputes; 2) cooperative, long-term efforts to improve the health of the Colorado River and its tributaries; and 3) development of additional water supply for those who live, work and play on the West Slope and for customers of Denver Water.

National Drought Mitigation Center. Innovative interstate watershed alliances are being developed to address long-term sustainability of water resources and health of riparian ecosystems in the face of uncertain social and environmental changes. In the Republican River Basin, Colorado, Nebraska, and Kansas have come together through seven resource conservation and development councils to create the Republican River Restoration Partnership (Knutson 2008). Through this partnership, they created the Republican River Basin Water and Drought Portal (National Drought Mitigation Center 2010) to provide stakeholders with tools for forecasting, climate and water information, planning and knowledge sharing.

The National Drought Mitigation Center, based at the University of Nebraska-Lincoln, works closely with stakeholders in the Great Plains region. The center conducted one study with over 160 local, state, tribal and federal water authorities to look at low-flow impacts in different areas of the Great Plains in an effort to develop a low-flow early warning system with the NOAA National Weather Service (Knutson et al. 2008). Studies have been carried out in the Upper Trinity River Basin in Texas, the Souris-Red River in North Dakota, and the Missouri River Basin. One outcome of these types of partnerships was the creation of a drought risk management website for ranchers (National Drought Mitigation Center 2012).

Western Governor's Association. Several multi-agency and/or multi-state climate change planning initiatives include states from the Great Plains. Many of these have been spearheaded through the Western Governors' Association (WGA) sustainability initiatives, which include the Water Needs and Strategies for a Sustainable Future program that began with reports in 2006 and 2008 and gave consensus recommendations for how the Western states should work with federal, local and private sector partners to address a range of issues. These issues include providing water supply to meet future demands, maintaining water supply infrastructure, resolving Indian water rights, preparing for climate change, and conserving endangered species (Western Governors' Association 2006, 2008). In the 2008 report, the initiative partners recommended the creation of WestFAST (Western Federal Agency Support Team; http://www.westernstateswater.org/westfast/) to assist states in implementing the reports' recommendations. This created a partnership between the Western States Water Council and eleven federal agencies that have water-resource responsibilities in the western US. The agencies created a work plan in 2011 to address three key areas: 1) climate change; 2) water availability, water use, and water reuse; and 3) water quality. To date, they have produced the WestFAST Water-Climate Change Program Inventory. Another outcome was the 2006 Shared Vision Partnership Agreement between the Western States Watershed Council and the US

Army Corps of Engineers that produced the Western States Watershed Study, demonstrating how federal agencies could work collaboratively with Western states on planning activities (U.S. Army Corps of Engineers 2009). Multiple state and federal agencies and entities were involved and the study adopted a shared vision, identified water data needs and gaps -- with federal and local water managers working together to evaluate new flood storage rule curves under a changing climate -- and enhanced federal interagency collaboration with the WSWS. Multiple additional goals and planning were identified for the continued collaboration.

In 2009, the Western Governors' Association adopted a policy resolution titled *Supporting the Integration of Climate Change Adaptation Science in the West* that created the Climate Adaptation Work Group, composed of Western state experts in air, forests, water and wildlife to recommend next steps. In 2010, the WGA) published its scoping *Climate Adaptation Priorities* report (Western Governors' Association 2010) which recommended increased collaboration and coordination among agencies and local stakeholders, support and sharing of appropriately scaled climate science and adaptation strategies, and an enhanced working relationship with Congress to educate members on the priorities for Western states and support needed for implementing adaptation (Western Governors' Association 2010).

Framework for Integration of Research, Analysis, Assessment, and Communication Activities: Steps Forward

In the Great Plains and throughout the nation, efforts to respond to climate change have emerged during the past decade. Strategies to reduce activities that contribute to climate change have been developed for a longer time and are further developed than those strategies dealing with adaptation to climate change. However, a growing recognition that adaptation strategies are needed has emerged in a number of sectors and communities around the world and the US (Wilby and Vaughan 2011). Being "climate smart" means implementing specific, measurable, achievable, realistic, and time-bound activities to reduce climate sensitivity and increase resilience to climate variability and change. Wilby and Vaughan (2011) identified nine hallmarks of organizations that are adapting to climate change, which include:

1. Visionary leadership
2. Objective setting
3. Risk and vulnerability assessment
4. Guidance for practitioners and research groups
5. Organizational learning
6. Low-regret adaptive management
7. Multi-partner working groups
8. Monitoring and reporting progress to inform adaptive management actions
9. Effective communications

The preceding section of this chapter provide a number of case studies of the ongoing efforts in the Great Plains. These case study examples are not a full compilation of efforts, but they represent a rich and varied set of activities, and in many ways, embrace the nine points identified above. What is apparent is that there are few well-coordinated efforts between agencies or institutions. In addition, it is difficult to learn about these efforts, and no clear mechanism to share knowledge of monitoring activities, impact analyses , climate information sources, or development of response strategies to climate change. This lack of coordination and communication results in a great inefficiency and limits the ability to access information on climate change impacts and focus research activities more strategically in the region.

Current efforts through NOAA/ RISA nodes and the Department of Interior Regional Climate Science Centers have been established to serve as a resource to regional efforts to provide better information on climate dynamics, impacts of climate changes, vulnerability and risk assessments, and how information can guide climate change responses across multiple sectors and supporting management and decision-making communities. These entities (RISAs and CSCs) are developing strategies to better coordinate among state and federal agencies and to provide a more comprehensive information portal where managers and decision makers can more readily find scientific information, including analysis of impacts and consequences to guide development of specific strategies to cope with climate change.

In general, risk is defined within the National Climate Assessment as the product of the likelihood of some climate impact plus the consequence of that event or climate stress. Global climate change projections help us to understand the range of possible future climates and the impacts of climate change to some degree. But on smaller, local or regional scales, there is considerable uncertainty of the time and spatial scales needed for decision making. Additionally, the possible future climate and its impacts on social-ecological systems depend on how we as a society adapt and mitigate climate change. Therefore, decisions we make now about how to plan for climate change are inherently uncertain.

The US National Climate Assessment risk framework is designed for scientific analysis, however, social science risk analysis and decision science shows that most risk decisions are made based on emotions and experience versus analytical processes or scientific evidence (Balstad et al. 2009). This is important to understand when linking probabilistic risk assessments with decision making. It accounts for the disconnect that sometimes occurs between what scientists think should be done and the reality of how decisions are made. Conversely, decision makers need to recognize this and strive to incorporate scientific findings in planning. However, there are times when analytical processes can predominate, especially when discussed in a group and when data and scientific information are clearly presented in a way that is relevant to the decision being made and the options being considered (Balstad et al. 2009). This calls for participatory research, iterative risk-based analysis between researchers and stakeholders, and collaborative decision making.

The information presented in this report is intended to help meet societal needs to respond to climate change. Research efforts at the various centers will be guided by user needs, in addition to scientific directions to better reduce or communicate more clearly the uncertainties in the information available. Engagement with managers and decision makers from a variety of sectors will be undertaken to ensure knowledge sharing between communities and researchers. The Great Plains is fortunate to have a number of highly respected centers, not only the RISAs or the CSCs, but also, the National Center for Atmospheric Research , NOAA Earth System Research Laboratory, High Plains Drought Center, National Drought Impact Science Center, EROS-Data Center, USFS Rocky Mountain Experimental Station, Department of Energy regional offices for Region 6, 7, and 8, the Department of Energy National Renewable Energy Laboratory, and many others.

There are lessons to be learned from efforts of the Western Governors' Association that illustrate how state, federal, tribal, and academic communities can work in a coordinated fashion to develop and implement strategies to deal with critical regional needs related to climate change. Issues, including water resources, land use, forest fire, and conservation, have been proactively addressed over the years, as mentioned above. The WGA has helped to define issues and to provide a framework to address these across the West. Other regional efforts include river basin initiatives, such as the Missouri River Basin efforts and the various agency coordination efforts to deal with flood control, land use practices, and conservation efforts. These bodies have a goal to provide better communication and, where needed, coordination of actions to deal with specific issues. The energy sector also has regional action groups, as mentioned in Chapter 6 of this report. However, assessments of climate change impacts and long-lasting climate change solutions need to be developed across sectors and include multiple stakeholders. We need to create a platform to support this more integrative effort in the research and the management activities implemented across the Great Plains.

The multi-agency approach of the US Global Change Research Program can help enable this coordinated effort across the region. However, real and lasting engagement with regional leaders and communities will be necessary to better assess stakeholders' vulnerability and adaptive capacity to deal with opportunities and challenges resulting from climate changes. Establishing a collaborative network responsible for communication between communities will greatly enhance the region's ability to respond to climate changes and to better create opportunities as changes unfold.

References

Adger, W. N., Agrawala, S., Mirza, M. M. Q., Conde, K., O'Brien, K., Pulhin, J., … Takahashi, K. (2007). Assessment of Adaptation Practices, Options, Constraints, and Capacity, in Climate Change 2007: Impacts, Adaptation and Vulnerability. In M. L. Parry, O. F. Canziani, J. P. Palutikof, P. J. van der Linden, & C. E. Hanson (Eds.), *Contributions of Working Group II to the Fourth Assessment of the Intergovernmental Panel on Climate Change* (pp. 717–743). Cambridge, U.K.: Cambridge University Press.

Adger, W. N., Brooks, N., Bentham, G., Agnew, M., & Eriksen, S. (2004). *New Indicators of Vulnerability and Adaptive Capacity, Technical Report* (p. 128). Norwich, England.

Adger, W. N., Brown, K., Nelson, D. R., Berkes, F., Eakin, H., Folke, C., … Tompkins, E. L. (2011). Resilience implications of policy responses to climate change. *WIREs Clim Change, 2*, 757–766.

Adger, W. N., Brown, K., & Tompkins, E. (2005). The political economy of cross-scale networks in resource co-management. *Ecology And Society, 10*(2), 9. Retrieved from http://www.ecologyandsociety.org/vol10/iss2/art9/

Adger, W. N., Eakin, H., & Winkels, A. (2009). Nested and teleconnected vulnerabilities to environmental change. *Front Ecol Environ, 7*(3), 150–157.

Adger, W. N., & Kelly, P. M. (1999). Social Vulnerability to Climate Change and the Architecture of Entitlements. *Mitigation and Adaptation Strategies for Global Change, 4*(3-4), 256–266.

Adler, P. R., Del Grosso, S. J., & Parton, W. J. (2007). Life-Cycle Assessment of Net Greenhouse-Gas Flux for Bioenergy Cropping Systems. *Ecological Applications, 17*(3), 675–691.

Ainsworth, E. A., & Long, S. . (2005). What have we leared from 15 years of free-air CO_2 enrichment (FACE)? A meta-analyic review of the responses of photosynthesis, canopy properties and plant production to rising CO_2. *New Phytologist, 165*, 351–372.

Akin, D. E., Fales, S. L., Rigsby, L. L., & Snook, M. E. (1987). Temperature effects on leaf anatomy, phenolic acids, and tissue digestibility in tall fescue. *Agronomy Journal, 79*, 271–275.

Albrecht, M. (2003, November 25). Progress made, but Fort Yates still dry. *The Bismark Tribune.*

Aldridge, G., Inouye, D. W., Forrest, J., Barr, W. A., & Miller-Rushing, A. J. (2011). Emergence of a mid-season period of low floral resources in a montane meadow ecosystem associated with climate change. *J.Ecol., 99*, 905–913.

Allen, C. D. et al. (2010). A global overview of drought and heat-induced tree mortality reveals emerging climate change risks for forests. *Forest Ecology and Management2, 259*, 660–684.

Alley, W. M. (2001). Ground water and climate. *Ground Water, 39*(2), 161.

Allison, H. E., & Hobbs, R. J. (2004). Resilience, adaptive capacity, and the "Lock-in Trap" of the Western Australian agricultural region. *Conservation Ecology, 9*(1).

American Society of Civil Engineers. (2009). *Report Card for America's Infrastructure* (p. March 25, 2009).

American Wind Energy Association. (2011). *U.S. Wind Industry Third Quarter Market Report* (p. 11). Washington, DC, USA.

Anderson, L. J., Maherali, H., Johnson, H. B., Polley, H. W., & Jackson, R. B. (2001). Gas exchange and photosynthetic acclimation over subambient to elevated CO2 in a C3–C4 grassland. *Global Change Biology, 7*, 693–707.

Anderson, M. T., Stamm, J. F., & Norton, P. A. (2008). Changes in Monthly Streamflow Conditions in the Missouri River Basin from 1957 to 2007: EOS, Transactions, American Geophysical Union, Fall Meeting Supplement, Abstract H11I-02. Retrieved from http://sd.water.usgs.gov/pubs/abstracts/Anderson2009MRNRCBiopForumAbstract.pdf

Anderson, R. C. (2006). Evolution and origin of the Central Grassland of North America: Climate, fire, and mammalian grazers. *Journal of the Torrey Botanical Society, 133*, 626–647.

Annie E. Casey Foundation. (2005a). Building More Secure Futures for Rural Families. *Casey Connects*, Winter.

Annie E. Casey Foundation. (2005b). *Strengthening Rural Families. Mortar and Muscle: Building Community and Assets through Self-Help Housing.*

Archer, S., Boutton, T. W., & Hibbard, K. A. (2001). Trees in grasslands: biogeochemical consequences of woody plant expansion. In E. D. Schulze, S. P. Harrison, M. Heimann, E. A. Holland, J. Lloyd, I. C. Prentice, & D. Schimel (Eds.), *Global Biogeochemical Cycles in the Climate System* (pp. 115–1337). San Diego, CA: Academic Press.

Arizmendi, D. (2003). Beyond Self-Help: A New Beginning for Colonias Residents. *Rural Voices, 8*(3), 21–22.

Aroonruengsawat, A., & Auffhammer, M. (2009). *Impacts of Climate Change on Residential Electricity Consumption: Evidence from Billing Data* (p. March). Retrieved from http://www.energy.ca.gov/2009publications/CEC-500-2009-018/CEC-500-2009-018-D.PDF

Augustine, N. H., Muggelstone, M. A., & Buckland, S. T. (1998). The role of simulation in modelling spatially correlated data. *Environmetrics, 9*, 175–196.

Averyt, K. J. (2011). The Coal Conundrum. In D. . Kenney & R. Wilkinson (Eds.), *The Water-Energy Nexus in the American West* (pp. 33–34). Northhampton, MA: Edward Elgar Publishing, Inc.

Averyt, K. J., Fisher, A. H. L., Lewis, A., Macknick, J., Madden, N., Rogers, J., & Tellinghuisen, S. (2011). *Freshwater use by U.S. power plants: Electricity's thirst for a precious resource. A report of the Energy and Water in a Warming World initiative* (p. 62). Cambridge, Massachusetts: Union of Concerned Scientists.

Axelrod, D. I. (1985). Rise of the grassland biome, central North America. *Botanical Review, 52*, 163–201.

Bailey, D. W. (2004). Management strategies for optimal grazing distribution and use of arid rangelands. *J. Anim. Sci, 82*(E Suppl.), E147–E153.

Baldocchi, D., & Wong, S. (2006). *White Paper: An Assessment of the Impacts of Future CO_2 and Climate on Californian Agriculture* (p. March).

Balstad, R., Russell, R., Gill, V., & Marx, S. (2009). *Adapting to an Uncertain Climate on the Great Plains: Testing Hypotheses on Historical Populations, in Adapting to Climate Change: Thresholds, Values, and Governance.* (N. W. Adger, I. Lorenzoni, & K. L. O'Brien, Eds.) (pp. 283–295). Cambridge, U.K.: Cambridge University Press.

Barnett, T. P., Adam, J. C., & Lettenmaier, D. P. (2005). Potential impacts of a warming climate on water availability in snow-dominated regions. *Nature, 438*(7066), 303–309.

Barnett, T. P., Pierce, D. W., Hidalgo, H. G., Bonfils, C., Santer, B. D., Das, T., … Dettinger, M. D. (2008). Human-induced changes in the hydrology of the Western United States. *Science, 319*(5866), 1080–1083.

Barnett, T., & Pierce, D. (2009). Sustainable water deliveries from the Colorado River in a changing climate. *Proceedings of the National Academy of Sciences, 106*(18), 7334.

Barney, J. N., & DiTomaso, J. M. (1977). Bioclimatic predictions of habitat suitability for the biofuel switchgrass in North America under current and future climate scenarios. *Biomass and Bioenergy, 8*, 255–283.

Barrett, C. B., Reardon, T., & Webb, P. (2001). Nonfarm income diversification and household livelihood strategies in rural Africa: concepts, dynamics, and policy implications. *Food Policy, 26*(4), 315–331.

Barry, R. G. (1983). Climatic environment of the Great Plains, Past and present. *Transactions of the Nebraska Academy of Science, XI*(Special Issue), 45–55.

Basin Electric Power Cooperative. (2011). *Ice storm devastates co-op systems*. Bismark, ND.

Bates, J. D., Svejcar, T., Miller, R. F., & Angell, R. A. (2006). The effects of precipitation timing on sagebrush steppe vegetation. *Journal of Arid Environments, 64*(670-391).

Baxter, R. M. (1977). Environmental effects of damns and impoudments. *Annual Reviews in Ecological Systems*, 255–283.

Beach, R. H., Adams, D. M., Alig, R. J., Baker, J. S., Latta, G. S., McCarl, B. A., … White, E. (2010). *Model Documentation for the Forest and Agricultural Sector Optimization Model with Greenhouse Gases (FASOMGHG)*. Washington, DC, USA.

Beauchemin, K. A., Janzen, H. H., Little, S. M., McAllister, T. A., & McGinn, S. . (2011). Mitigation of green house gas emissions from beef production in Western Canada – Evaluation using farm-based life cycle assessment. *Anim. Feed. Sci. Technol., 166-167*, 633–677.

Beauchemin, K., Janzen, H. H., Little, S. M., McAllister, T. A., & McGinn, S. M. (2010). Life cycle assessment of greenhouse gas emission from beef production in western Canada: A case study. *Agric. Syst., 103*, 371–379.

Becker, C. G., Fonseca, C. B., Haddad, C. F. B., Batista, R. F., & Prado, P. I. (2007). Habitat split and the global decline of amphibians. *Science, 318*, 1775–1777.

Bell, C., & Taylor, J. (2008). *Water Laws and Policies for a Sustainable Future: A Western States' Perspective* (p. June). Murray, UT.

Bell, M. L., Dominici, F., & Samet, J. M. (2005a). A meta-analysis of time-series studies of ozone and mortality with comparison to the national morbidity, mortality, and air pollution study. *Epidemiology, 16*(4), 436–445.

Bell, M. L., Dominici, F., & Samet, J. M. (2005b). A meta-analysis of time-series studies of ozone and mortality with comparison to the national morbidity, mortality, and air pollution study. *Epidemiology, 16*(4), 436–445.

Benedict, R. A., Freeman, P. W., & Genoways, H. H. (1996). Prairie legacies—mammals. In F. B. Samson & F. L. Knopf (Eds.), *Prairie Conservation: Preserving North America's Most Endangered Ecosystem* (pp. 149–166). Covelo, CA: Island Press.

Benke, A. ., & Cushing, C. E. (2005). *Rivers of North America*. (A. . Benke & C. E. Cushing, Eds.). New York, NY: Elsevier Academic Press.

Berkes, F. (2004). Rethinking Community-Based Conservation. *Conservation Biology, 18*(3), 621–630.

Berkes, F., & Folke, C. (1998). *Linking Social and Ecological Systems: Management Practices and Social Mechanisms for Building Resilience* (p. 459). Cambridge, U.K.: Press Syndicate of the University of Cambridge.

Bestelmeyer, B. T., Brown, J. R., Fuhlendorf, S. D., Fults, G. A., & Wu, X. B. (2011). A landscape approach to rangeland conservation practices. In D. D. Briske (Ed.), *Conservation Benefits of Rangeland Practices: Assessment, Recommendations and Knowledge Gaps* (pp. 337–370). Washington, DC, USA: USDA Natural Resources Conservation Service.

Birkmann, J., Garschagen, M., Kraas, F., & Quang, N. (2010). Adaptive Urban Governance - New Challenges for the Second Generation of Urban Adaptation Strategies to Climate Change. *Sustainability Science, 5*(2), 185–206.

Bjune, A. E., Sepp¨a, H., & Birks, H. J. B. (2009). Quantitative summer temperature reconstructions for the last 2000 years based on pollen-stratigraphical data from northern Fennoscandia. *J. Paleolimnol., 41*, 43–56.

Black, C. C. (1973). Photosynthetic carbon fixation in relation to net CO2 uptake. *Annual Rev Plan Physiol, 24*, 253–286.

Blanco-Canqui, H., Schlegel, A. J., & Heer, W. F. (2011). Soil-profile distribution of carbon and associated properties in no-till along a precipitation gradient in the central Great Plains. *Agriculture, Ecosystems and Environment, 144*(1), 107–116.

Bolen, E. G., Smith, L. M., & Schramm, H. L. (1989). Playa lakes: prairie wetlands of the Southern High Plains. *BioScience, 39*, 615–623.

Bonan, G. B., Levin, S., Kergoat, L., & Oleson, K. W. (2002). Landscapes as patches of plant functional types: An integrating concept for climate and ecosystem models. *Global Biogeochemical Cycles, 16*(10), 1029–2000.

Bond, W. J. (2008). What Limits Trees in C-4 Grasslands and Savannas? *Annual Review of Ecology and Systematics, 39*(641-359).

Bonner, T. H., & Wilde, G. R. (2000). Changes in teh Canadian River fish assemblage associated with reservoir construction. *Journal of Freshwater Ecology, 15*, 189–198.

Bosch, O. J. H., Ross, A. H., & Beeton, R. J. S. (2003). Integrating Science and Management Through Collaborative Learning and Better Information Management. *Systems Research and Behavioral Science, 20*(2), 107–118.

Bottrell, C. E., Ingersol, R. H., & Jones, R. W. (1964). Notes on the embryology, early development, and behavior of Hybopsis aestivalis tetranemus (Gilbert). *Trans. Amer. Microscopy Soc., 83*, 391–399.

Bragg, T. B., & Hulbert, L. C. (1976). Woody plant invasion of unburned Lansas bluestem prairie. *Journal of Range Management, 29*, 19–24.

Brekke, L. D., Kiang, J. E., Olsen, J. R., Pulwarty, R. S., Raff, D. A., Turnipseed, D. P., ... White, K. D. (2009). *Climate change and water resources management—A federal perspective: U.S. Geological Survey Circular 1331* (p. 65).

Briechle, K. J. (2009). *Kansas City Green Infrastructure Case Study #10, in Green Infrastructure Case Study Series* (p. 14). Arlington, VA.

Briggs, J. M., Knapp, A. K., Blair, J. M., Heisler-White, J. L., Hoch, G. A., M.S., L., & McCarron, J. K. (2005). An ecosystem in transition. Causes and consequences of the conversion of mesic grassland to shrubland. *BioScience, 55*, 243–254.

Brikowski. (2008). Doomed reservoirs in Kansas, USA? Climate change and groundwater mining on the Great Plains lead to unsustainable surface water storage. *Journal of Hydrology, 354*, 90–101.

Briske, D. D., Jolley, L. W., Duriancik, L. F., & Dobrowolski, J. P. (2011). Introduction to the Conservation Effects Assessment Project and the rangeland literature synthesis. In D. D. Briske (Ed.), *Conservation Benefits of Rangeland Practices: Assessment, Recommendations and Knowledge Gaps* (pp. 1–8). Washington, DC, USA: USDA Natural Resources Conservation Service.

Burris, L. E., & Skagen, S. K. (2012). Modeling sediment accumulation in North American playa wetlands in response to climate change, 1940-2100. *Climatic Change*, 1–15.

Cawood, M. (2011). Cows conserve carbon. *Feedstuffs Magazine, 83*(42), 5.

Centers for Disease Control and Prevention, U.S. Environmental Protection Agency, NOAA, & American Water Works Association. (2010). *When every drop counts: protecting public health during drought conditions – a guide for public health professionals. 2010.* Atlanta, GA.

Changnon, S., Kunkel, K., & Reinke, B. (1996). Impacts and responses to the 1995 heat wave: A call to action. *Bull. Amer. Meteor. Soc., 77*, 1497–1506.

Chase, J. M., & Knight, T. M. (2003). Drought-induced mosquito outbreaks in wetlands. *Ecology Letters, 6*, 1017–1024.

Chase, K. J. (2011). *Development of a precipitation-runoff model to simulate unregulated streamflow in the South Fork Flat-head River Basin, Montana: U.S. Geological Survey Scien¬tific Investigations Report 2011–5095* (p. 39).

Chen, I.-C., Hill, J. K., Ohlemüller, R., Roy, D. B., & Thomas, C. D. (2011). Rapid Range Shifts of Species Associated with High Levels of Climate Warming. *Science, 333*(1024-1026).

Chiu, Y., Walseth, B., & Suh, S. (2009). Water footprint of biofuel in the U.S. *Environmental Science and Technology, 43*(8), 2688–2692.

Christensen, J. H., Hewitsonb, B., Busuioc, A., Chen, A., Gao, X., Held, I., … Whetton, P. (2007). Regional Climate Projections. In S. Solomon, D. Qin, M. Manning, Z. Chen, M. Marquis, t K. B. Avery, … H. L. Miller (Eds.), *Climate Change 2007: The Physical Science Basis. Contribution of Working Group I to the Fourth Assessment Report of the Intergovernmental Panel on Climate Change.* Cambridge, U.K. and New York, NY, USA: Cambridge University Press.

City of Dallas. (2010). *Water Conservation Five-Year Strategic Plan* (p. Updated June 2010). Retrieved from http://savedallaswater.com/pdf/ConservationStrategicPlan.pdf

City of Grand Forks. (2012). *Greenway Fact Sheet - Historic Flood- Remarkable Recovery, Grand Forks, ND, 10 Years Later…Rebuilt, Renewed, Reborn.*

City of Greensburg. (2008). *Greensburg Sustainable Comprehensive Plan.* Retrieved from http://www.greensburgks.org/residents/recovery-planning

City of Houston. (2012). Public Works and Engineering: Drinking water operations. Retrieved January 03, 2012, from http://www.publicworks.houstontx.gov/utilities/drinkingwater.html

Clark, C. . (2007). *Consequences and reversibility of community and biogeochemical shifts following long-term N addition. Dissertation.* St. Paul, MN, USA.

Clark, C. ., Harto, C. B., Sullivan, J. L., & Wang, M. Q. (2011). *Water use in the development and operation of geothermal power plants.* Argonne, IL.

Classen, R., Carriazo, F., Cooper, J. C., Hellerstein, D., & Udea, K. (2011). *Grassland to Cropland Conversion in the Northern Plains: The Role of Crop Insurance, Commodity, and Disaster Programs, ERR-120.*

Cleland, E. E., Chiariello, N. R., Loarie, S. R., Mooney, H, A., & Field, C. B. (2006). Diverse responses of phenology to global changes in a grassland ecosystem. *Proceedings of the National Academy of Sciences, 103*, 13740–13744.

Climatologist, O. of the T. S. (2011). Texas Drought Officially the Worst Ever. Press Release. August 4, 2011. *College of Geosciences, Department of Atmospheric Sciences, Texas A&M University.* Retrieved June 02, 2012, from http://tamunews.tamu.edu/2011/08/04/texas-drought-officially-the-worst-ever/

Coakley, S. M., Scherm, H., & Chakraborty, S. (1999). Climate change and plant disease management. *Annual Review of Phytopathology, 37*, 399–426.

COGCC. (2012). *Water Sources and Demand for the Hydraulic Fracturing of Oil and Gas Wells in Colorado from 2010 through 2015* (p. 9).

Colby, B. G., & Tanimoto, P. (2011). Using Climate Information to Improve Electric Utility Load Forecasting. In Adaptation and Resilience: The Economics of Climate, Water and Energy Challenges. In B. G. Colby & G. B. Frisvold (Eds.), *The American Southwest* (pp. 207–228). Washington, DC, USA: RFF Press.

Colby, B. G., Thorson, J. E., & Britton, S. (2005). *Negotiating Tribal Water Rights – Fulfilling Promises in the Arid West*. University of Arizona Press.

Collins, S. L., Knapp, A. K., Briggs, J. M., Blair, J. M., & Steinauer, E. M. (1998). Modulation of diversity by grazing and mowing in native tallgrass prairie. *Science, 280*, 745–747.

Cook, B. I., Miller, R. L., & Seager, R. (2009). Amplification of the North American "Dust Bowl" drought through human-induced land degradation. *Proceedings of the National Academy of Sciences, 106*, 4997–5001.

Cooley, H., Fulton, J., & Gleick, P. H. (2011). *Water for Energy: Future Water Needs for Electricity in the Intermountain West. 64*. Oakland, CA.

Corn, P. S. (2005). Climate change and amphibians. *Animal Biodiversity and Conservation, 28*, 59–67.

Corn, P. S., & Peterson, C. R. (1996). Prairie legacies—amphibians and reptiles. In F. B. Samson & F. L. Knopf (Eds.), *Prairie Conservation: Preserving North America's Most Endangered Ecosystem* (pp. 125–134). Covelo, CA: Island Press.

Costanza, R., D'Arge, R., DeGroot, R., Farber, S., Grasso, M., Hannon, B., … Belt, M. V. D. (1997). The value of the world's ecosystem services and natural capital. *Nature, 387*, 253–260.

Covich, A. P., Fritz, S. C., Lamb, P. J., Marzolf, R. ., Mathews, W. J., Poiani, K. A., … Winter, T. C. (1997). Potential effects of climate change on aquatic ecosystems of the Great Plains of North America. *Hydrol. Process, 11*, 993–1021.

Coviella, C., & Trumble, J. (1999). Effects of elevated atmospheric carbon dioxide on insect-plant interactions. *Conservation Biology, 13*, 700–712.

Craine, J. M., Spurr, R., McLauchlan, K. K., & Fierer, N. (2010). Landscape-level variation in temperature sensitivity of soil organic carbon decomposition. *Soil Biology and Biochemistry, 42*, 373–375.

Craine, J. M., Wolkovich, E. M., Towne, E. G., & Kembel, S. W. (2011). Flowering phenology as a functional trait in a tallgrass prairie. *New Phytologist, 193*, 673–382.

Croissant, R. L., Peterson, G. A., & Westfall, D. G. (2008). *Dryland Cropping Systems: Fact Sheet No. 0.516. Colorado State University Extension Crop Series Factsheet No. 0.516* (pp. 11–13). Retrieved from http://www.ext.colostate.edu/pubs/crops/00516.pdf

Cross, F. B., Moss, R. E., & Collins, J. T. (1985). *Assessment of dewatering impacts on stream fisheries in the Arkansas and Cimarron Rivers. Nongame Wildlife Contract No. 46.*

Cully, A. C., Cully, J. F. J., & Hiebert, R. D. (2003). Invasion of exotic plant species in tallgrass prairie fragments. *Conservation Biology, 17*, 990–998.

D., D., & Lewis, G. (2008). Rio Grande Salinity Management – A Real Possibility? *Southwest Hydrology, March/Apri*. Retrieved from http://www.swhydro.arizona.edu/archive/V7_N2/feature5.pdf

Dahl, T. E. (1990). *Wetlands Losses in the United States: 1780s to 1980s*. Washington, DC, USA.

Daily, G. C. (1997). *Nature's Services: Societal Dependence on Natural Ecosystems*. Washington, DC, USA: Island Press.

Daily, G. C., & Ellison, K. (2002). *The New Economy of Nature: The Quest to Make Conservation Profitable*. Washington, DC, USA: Island Press.

Dale, V. H. (1997). The relationship between land-use change and climate change. *Ecological Applications, 7*, 753–769.

Dale, V. H., Efroymson, R. A., & Kline, K. L. (2011). The land use—climate change—energy nexus. *Landscape Ecology, 26*, 755–773.

Dallas, C. of. (2012). Dallas Water Utilities: Water Utilities FAQ. Retrieved January 03, 2012, from http://www.dallascityhall.com/dwu/water_utilities_faqs3.html

Davidson, E. A., & Ackerman, I. L. (1993). Changes in soil carbon inventories following cultivation of previously untilled soils. *Biogeochemistry, 20*, 161–193.

Dawson, B. (2012). *A Drought for the Centuries: It Hasn't Been This Dry in Texas Since 1789* (p. February 8, 2012). Retrieved from http://texasclimatenews.org/wp/?p=3355

De Boeck, H. J., Liberloo, M., Gielen, B., Nijs, I., & Ceulemans, R. (2008). The observer effect in plant science. *New Phytologist, 177*, 579–583.

De Steiguer, J. E., Brown, J. R., & Thorpe, J. (2008). Contributing to the mitigation of climate change using rangeland management. *Rangelands, 30*(3), 7–11.

Debinski, D. M., Wickham, H., Kindscher, K., Caruthers, J. C., & Germino, M. (2010). Montane meadow change during drought varies with background hydrologic regime and plant functional group. *Ecology, 91*, 1672–1681.

Delpa, I., Baures, A., Jung, V., & Thomas, O. (2009). Impacts of climate change on surface water quality in relation to drinking water production. *Environment International, 35*, 1225–1233.

Dennehy, K. F. (2000). *High Plains regional ground-water study. USGS Fact Sheet FS–091–00.* Reston, Virginia.

Denver Mile High Million Initiative. (2012). *Cool Schools Tree Planting program* (p. February 20, 2012). Retrieved from http://milehighmillion.org/pages/section/programs/cool-schools-tree-planting-program

Denver Water. (2011a). Recycled Water for Soil and Trees Fact Sheet. Retrieved from http://www.denverwater.org/WaterQuality/RecycledWater/RecycledWaterNews/

Denver Water. (2011b). *Solutions:Saving Water For the Future.* Denver, CO. Retrieved from http://www.denverwater.org/docs/assets/44EC9CA1-A444-8348-68E2A8D5BD36DFA4/Solutions2011.pdf

Denver Water. (2012). *Planning for an Uncertain Future.*

Dessai, S., Lu, X. F., & Risbey, J. S. (2005). On the role of climate scenarios for adaptation planning, Glob. Environ. *Change-Human Policy Dimens., 15*(2), 87–97.

Dijkstra, F. A., Blumenthal, D., Morgan, J. A., Pendall, E., Carrillo, Y., & Follett, R. F. (2010). Contrasting effects of elevated CO2 and warming on nitrogen cycling in a semiarid grassland. *New Phytologist, 187*, 426–437.

Dobson, A. P., & Foufopoulos, J. (2011). Emerging infectious pathogens in wildlife. *Philosophical Transactions of the Royal Society B, 356*, 1001–1012.

Dodds, W. K., Gido, K., Whiles, M. R., Fritz, K. M., & Matthews, W. J. (2004). Life on the edge: the ecology of Great Plains prairie streams. *BioScience, 54*, 205–216.

Dolan, A. H., & Walker, I. J. (2006). Understanding vulnerability of coastal communities to climate change related risks. *J Coastal Res, 3*, 1316–1323.

Douglas, M., & Wildavskky, A. (1982). *Risk and Culture: An Essay on the Selection of Technical and Environmental Dangers.* Berkeley, CA: University of California Press.

Drawbaugh, K. (2011). *Insurers Fight Over Flood Policies Dropped by State Farm* (p. May 6, 2011). Retrieved from http://www.insurancejournal.com/news/national/2011/05/06/197439.htm

Drummond, M. A., Auch, R. F., Karstensen, K. A., Sayler, K. L., Taylor, J. L., & Loveland, T. R. (2012). Land change variability and human-environment dynamics in the United States Great Plains. *Land Use Policy, 29*(3), 710–723.

Duncan, M., Fisher, D., & Drabenstott, M. (1995). Planning for a sustainable future in the Great Plains. In D. A. Whilhite, D. A. Wood, & K. H. Smith (Eds.), *Proceedings of the Symposium – Planning for a sustainable Future: The Case of the North American Great Plains* (pp. 23–42).

Dunnell, K. L., & Travers, S. E. (2011). Shifts in the flowering phenology of the northern Great Plains: Patterns over 100 years. *Am. J. Bot.*, *98*, 935–945.

Durham, B. W., & Wilde, G. R. (2006). Influence of stream discharge on reproductive success of a prairie stream fish assemblage. *Transactions of the American Fisheries Society*, *135*, 1644–1653.

Durham, B. W., & Wilde, G. R. (2009a). Effects of streamflow and intermittency on the reproductive success of two broadcast-spawning cyprinid fishes. *Copeia*, *2009*, 21–28.

Durham, B. W., & Wilde, G. R. (2009b). Population dynamics of small-eye shiner an imperiled cyprinid fish endemic to the Brazos River, Texas. *Transactions of the American Fisheries Society*, *138*, 666–674.

Eakin, H. C., & Patt, A. (2011). Are adaptation studies effective, and what can enhance their practical impact? *Wiley Interdisciplinary Reviews: Climate Change*, *2*(2), 141–153.

East River Electric. (2006). *Growth & Storm: top challenges for local co-ops in 2005.*

Ebi, K. L., & Mcgregor, G. (2008). Climate change, tropospheric ozone and particulate matter, and health impacts. *Environmental Health Perspectives*, *116*(11), 1449–1455.

Ehleringer, J. R., & Bjorkman, O. (1977). Quantum yields for CO2 uptake in C3 and C4 plants: dependence on temperature, CO2 and 02 concentrations. *Plant Physiology*, *59*, 86–90.

Elam, T. E. (2007). Is Organic Beef and Dairy Production a Responsible Use of Our Resources ? In *22nd Southwest Nutrition and Management Conference* (pp. 66–75). Tempe, AZ. Retrieved from http://ag.arizona.edu/ans/swnmc/Proceedings/2007/Elam_2007SWNMC.pdf

Electric Reliability Council of Texas. (2012). ERCOT Quick Facts. Retrieved from http://ercot.com/content/news/presentations/2012/ERCOT Quick Facts - Jan 2012.pdf

Elgaali, E., & Garcia, L. A. (2007). Using Neural Networks to Model the Impacts of Climate Change on Water Supplies. *Journal of Water Resources Planning and Management*, *133*(3), 230.

Elias, S. A. (1991). Insects and climate change. *BioScience*, *41*, 552–559.

Ellis, S. ., Doerfer, J. T., Mustard, M. H., Blakely, S. R., & Gibbs, J. W. (1984). *Analysis of urban storm-runoff data and the effects on the South Platte River, Denver Metropolitan Area, Colorado. Water Resources Investigations Report 84-4159* (p. 66). Lakewood, CO.

Emslie, S. D. (2007). Fossil passerines from the early Pliocene of Kansas and the evolution of songbirds in North America. *Auk*, *124*(1), 85–95.

Encyclopedia Britannica Online. (2012). Arkansas River Navigation System. Retrieved from http://www.britannica.com/EBchecked/topic/34918/Arkansas-River-Navigation-System

Energy, A. (2011). Power Saver Program – Saving Energy Together Commercial Rebates. Ceiling Roof Insulation and Reflective Roof Coating. Retrieved from http://www.austinenergy.com/energy efficiency/Programs/Rebates/Commercial/Commercial Energy/lightingQualifications.pdf

Energy Information Administration. (2000). *Energy Consumption and Renewable Energy Development Potential on Indian Lands* (p. April 2000). Retrieved from ftp://ftp.eia.doe.gov/pub/pdf/renewables/ilands.pdf

Epstein, H. E., Burke, I. C., & Lauenroth, W. K. (2002). Regional patterns of decomposition and primary production rates in the U.S. Great Plains. *Ecology*, *83*, 320–327.

Epstein, P. R. (1998). *Health, Ecological, and Economic Dimensions of Global Change (HEED). Marine Ecosystems: Emerging Diseases as Indicators of Change. Marine Ecosystems: Emerging Diseases as Indicators of Change.* Boston, Massachusetts.

Epstein, P. R. (2001). Climate change and emerging infectious diseases. *Microbes Infect., 3,* 747–754.

ERCOT - Electric Reliability Council of Texas. (2011). *Grid Operations and Planning Report.* Retrieved from http://www.ercot.com/content/meetings/board/keydocs/2011/1212/Item_06e_-_Grid_Operations_and_Planning_Report.pdf

Eriksen, S. E. H., Klein, R. J. T., Ulsrud, K., Naess, L. O., & O'Brien, K. (2007). *Climate Change Adaptation and Poverty Reduction: Key Interactions and Critical Measures. Report prepared for the Norwegian Agency for Development Cooperation (Norad). GECHS Report 2007:1Rep.* Oslo, Norway.

Eriksen, S. H. (2011). When not every response to climate change is a good one: Identifying principles for sustainable adaptation. *Climate and Development, 3,* 7–20.

Eriksen, S. H., & Brown, K. (2011). Sustainable adaptation to climate change. *Climate and Development, 3,* 3–6.

Eriksen, S. H., & O'Brien, K. (2007). Vulnerability, poverty and the need for sustainable adaptation measures. *Climate Policy, 7*(4), 337–352.

Etkin, D., & Ho, E. (2007). Climate change: perceptions and discourses of risk. *J Risk Res, 10*(5), 623–641.

Euliss, N. H., J., Wrubleski, D. A., & Mushet, D. M. (1999). Invertebrates in wetlands of the Prairie Pothole Region: species composition, ecology and management. In D. Batzer, R. B. Rader, & S. A. Wissinger (Eds.), *Invertebrates in freshwater wetlands of North America: ecology and management* (pp. 471–514). New York, NY: John Wiley and Sons.

Fahrig, L. (2003). Effects of habitat fragmentation on biodiversity. *Annual Review of Ecology, Evolution, and Systematics, 34,* 487–515.

Fan, S., Methaprayoon, K., & Lee, W. J. (2007). Short-term Multi-Region Load Forecasting Based on Weather and Load Diversity Analysis. In *2007 39th North American Power Symposium (NAPS 2007).* Retrieved from http://ieeexplore.ieee.org/stamp/stamp.jsp?arnumber=4402366

Farag, A. T., Radwan, A. H., Sorour, F., El Okazy, A., El-Agamy, E., & El-Sebae, A. R. (2010). Chlorpyrifos induced reproductive toxicity in male mice. *Reproductive Toxicology, 29,* 80–85.

Farrelly, M. A., & Brown, R. R. (2011). Rethinking urban water management: experimentation as a way forward? *Global Environmental Change, 21*(2), 721–732.

Fay, P. A., Blair, J. M., Smith, M. D., Nippert, J. B., Carlisle, J. D., & Knapp, A. K. (2011). Relative effects of precipitation variability and warming on grassland ecosystem function. *Biogeosciences Discussion, 8,* 6859–6900.

Fay, P. A., Kaufman, D. M., Nippert, J. B., Carlisle, J. D., & Harper, C. W. (2008). Changes in grassland ecosystem function due to extreme rainfall events: Implications for responses to climate change. *Global Change Biology, 14,* 1600–1608.

FEMA. (2012). *Declared Disasters by Year or State Federal Emergency Management Agency.*

Field, C. B., Jackson, R. B., & Mooney, H. A. (1995). Stomatal Responses to Increased CO2 – Implications from the Plant to the Global-Scale. *Plant Cell and Environment, 18*(10), 1414–1225.

Field, C. B., Lobell, D. B., Peters, H. A., & Chiariello, N. R. (2007). Feedbacks of terrestrial ecosystems to climate change. *Annual Review of Environment and Resources, 32,* 1–29.

Fisher, J., & Ackerman, F. (2011). *The Water-Energy Nexus in the Western States: Projections to 2100* (p. 39). Somerville, MA. Retrieved from http://sei-us.org/Publications_PDF/SEI-Western-Water-Energy-0211.pdf

Fisher, J. R., & Paukert, C. P. (2008). Historical and current environmental influences on an endemic Great Plains fish. *Am. Midl. Nat., 159,* 364–377.

Flenner, I., & Sahlén, G. (2008). Dragonfly community re-organisation in boreal forest lakes: rapid species turnover driven by climate change? *Insect Conservation and Diversity, 1*, 169–179.

Folke, C., Hahn, T., Olsson, P., & Norberg, J. (2005). Adaptive Governance of Social-Ecological Systems. *Annual Review of Environment and Resources, 30*(1), 441–473. doi:10.1146/annurev.energy.30.050504.144511

Ford, J. D., Keskitalo, E. C. H., Smith, T., Pearce, T., Berrang-Ford, L., Duerden, F., & Smit, B. (2010). Case study and analogue methodologies in climate change vulnerability research. *Wiley Interdisciplinary Reviews: Climate Change, 1*(3), 374–392.

Forman, S. L., Oglesby, R., & Webb, R. S. (2001). Temporal and spatial patterns of Holocene dune activity on the Great Plains of North America: megadroughts and climate links. *Global and Planetary Change, 29*, 1–29.

Foster, J., Lowe, A., & Winkelman, S. (2011). *The Center for Clean Air Policy. The Value of Green Infrastructure for Urban Adaptation* (p. February 2011).

Foti, R., Ramierz, J. A., & Brown, T. C. (2011). *Vulnerability of U.S. Water Supply to Shortage Draft General Technical Report*. Fort Collins, CO.

Franzluebbers, A. J., Steiner, J. L., Karlen, D., Griffin, T., Singer, J., & Tanaka, D. (2011). Rainfed Farming Systems in the USA. In P. et al. Tow (Ed.), *Rainfed Farming Systems* (pp. 511–560). Springer Science Business Media.

Friedel, M. H. (1991). Range condition assessment and the concept of thresholds: a viewpoint. *Journal of Range Management, 44*, 422–426.

Fussel, H. M. (2007). Adaptation Planning for Climate Change: Concepts, Assessment Approaches, and Key Lessons. *Sustainability Science, 2*(2), 265–275.

Fussel, H. M., & Klein, R. J. T. (2006). Climate Change Vulnerability Assessments: An Evolution of Conceptual Thinking. *Climatic Change, 75*(3), 301–329.

G.A., P., A.D., H., J.L., H., O.R., J., D.J., L., & D.L., T. (1998). Reduced tillage and increasing cropping intensity in the Great Plains conserves soil C. *Soil and Tillage Research, 47*(3-4), 207–218.

Galyean, M. L., Ponce, C., & Schutz, J. (2011). The future of beef production in North America. *Animal Frontiers, 1*(2), 29–36. doi:10.2527/af.2011-0013

Ganesan, G. (2010). *Estimating recharge through playa lakes to the southern High Plains aquifer. MSc Thesis* (p. 80). Lubbock, TX.

Garrett, K. A., Nita, M., DeWolf, E. D., Gomez, L., & Sparks, A. H. (2009). Plant pathogens as indicators of climate change. In T. Letcher (Ed.), *Climate change: observed impacts on planet Earth* (pp. 425–437). Dordrecht, Netherlands: Elsevier.

Garry, P. M., Spurlin, C., & Nelsen, D. (2009). Wind Energy in Indian Country: A Study of thee Challenges and Opportunities Facing South Dakota Tribes. *South Dakota Law Review, 54 SDL Rev*, 448.

Gates, T. K., Garcia, L. A., & Labadie, J. W. (2006). *Toward Optimal Water Management in Colorado's Lower Arkansas River Valley: Monitoring and Modeling to Enhance Agriculture and Environment. Colorado Water Resources Research Institute Completion Report No. 205. Colorado Agricultural Experiment Station Tech* (p. June 2006).

Gedney, N., Cox, P. M., Betts, R. A., Boucher, O., Huntingford, C., & Stott, P. A. (2006). Detection of a direct carbon dioxide effect in continental river runoff records. *Nature, 439*(7078), 835–838.

General Accounting Office. (2003). *Freshwater Supply: States' Views of How Federal Agencies Could Help Them Meet the Challenges of Expected Shortages. GAO-03-514*. Washington, DC, USA. Retrieved from http://www.gao.gov/new.items/d03514.pdf

Gentile, R., Vanlauwe, B., & Six, J. (2011). Litter quality impacts short- but not long-term soil carbon dynamics in soil aggregate fractions. *Ecological Applications, 21*, 695–703.

Georgescu, M., Lobell, D. B., & Field, C. B. (2011). Direct cliamte effects of perennial bioenergy crops in the United States. *Proceedings of the National Academy of Sciences, 10*, 4307–4312.

Gerdes, G. H. (2004). Rift Valley fever. *REview Scientifique et Technique, 23*, 613–623.

Gibbons, J. W., Scott, D. E., Ryan, T. J., Bulhmann, K. A., Tuberville, T. D., Metts, B. S., ... Winne, C. T. (2000). The global decline of reptils, déjà vu amphibians. *BioScience, 50*, 653–666.

Gido, K. B., Perkin, J. S., Johnson, E., & Tabor, V. M. (2010). *Consequences of Stream Fragmentation and Climate Change for Rare Great Plains Fishes. Kansas State University Report to the Great Plains Landscape Conservation Cooperative.* (p. 35). Retrieved from www.gplcc.org

Giusti, C. (2002). Nuestra Casa (Our House): A new model for self-help and improvement along the Texas/Mexico Border. *Texas Business Review, August.*

Glantz, M. H. (1999). *Creeping Environmental Problems and Sustainable Development in the Aral Sea Basin* (p. 291). New York, NY: Cambridge University Press.

Gleason, R. A., Euliss, N. H., Hubbard, D. E., & Duffy, W. B. (2003). Effects of sediment load on emergence of aquatic invertebrates and plants from wetland soil egg and seed banks. *Wetlands, 23*(26-34).

Gleason, R. A., Tangen, B. A., Browne, B. A., & Euliss, N. H. (2009). Greenhouse gas flux from cropland and restored wetlands in the Prairie Pothole Region. *Soil Biology and Biochemistry, 41*(2501-2507).

Glick, P., Stein, B. A., & Edelson, N. A. (2011). *Scanning the Conservation Horizon: A Guide to Climate Change Vulnerability Assessment*. Washington, DC, USA.

Gough, B. (2002). Panel V: Revitalizing Economies, Preserving Cultures, and Protecting the Environment: Striking the Balance in South Dakota and Indian Country. *Great Plains Nat. Resources J., 7*, 67.

Gough, L., Osenberg, C. W., Gross, K. L., & Collins, S. L. (2000). Fertilization effects on species density and primary productivity in herbaceous plant communities. *Oikos, 89*, 428–439.

Grainger, C., & Beauchemin, K. A. (2011). Can enteric methane emission from ruminants be lowered without lowering their production? *Anim. Feed. Sci. Technol., 166-167*, 308–320.

Graumlich, L. J., Pisaric, M. F. J., Waggoner, L. A., Littell, J. S., & King, J. C. (2003). Upper Yellowstone River flow and teleconnections with Pacific Basin climate variability during the past three centuries. *Climatic Change, 59*, 245–262.

Gray, M. J., Smith, L. M., & Leyva, R. I. (2004). Influence of agricultural landscape structure on a Southern High Plains amphibian assemblage. *Landscape Ecology, 19*, 719–729.

Green, T. R., Bates, B. C., Charles, S. P., & Fleming, P. M. (2007). Physically based simulation of potential effects of carbon dioxide–altered climates on groundwater recharge. *Vadose Zone J., 6*, 597–609.

Green, T. R., Taniguchi, M., Kooi, H., Gurdak, J. J., Allen, D. M., & Hiscock, K. M. et al. (2011). Beneath the surface of global change: impacts of climate change on groundwater. *Journal of Hydrology, 405*, 532–560.

Grimm, N. B., Foster, D., & Groffman, P. et al. (2008). The changing landscape: ecosystem responses to urbanization and pollution across climatic and societal gradients. *Front Ecol Environ, 6*, 264–72.

Gunderson, L. H., & Holling, C. S. (2002). *Panarchy: Understanding Transformations in Human and Natural Systems* (p. 507). Washington, DC, USA: Island Press.

Gunderson, L. H., Holling, C. S., & Light, S. (1995). *Barriers and bridges to renewal of ecosystems and institutions*. New York, NY.

Gurdak, J. J., Hanson, R. T., McMahon, P. B., Bruce, B. W., McCray, J. E., Thyne, G. D., & Reedy, R. C. (2007). Climate variability controls on unsaturated water and chemical movement, High Plains aquifer, USA. *Vadose Zone J., 6*, 533–547.

Gurdak, J. J., McMahon, P. B., & Bruce, B. W. (2011). Vulnerability of groundwater quality to human activity and climate change and variability, High Plains aquifer, USA. In H. Treidel, J. J. Martin-Bordes, & J. J. Gurdak (Eds.), *limate change effects on groundwater resources: A global synthesis of findings and recommendations* (p. 414). IAH - International Contributions to Hydrogeology. Taylor and Francis publishing.

Gurdak, J. J., & Roe, C. D. (2009). *Recharge Rates and Chemistry Beneath Playas of the High Plains Aquifer - A Literature Review and Synthesis: U.S. Geological Survey Circular 1333* (p. 39). Retrieved from http://pubs.usgs.gov/circ/1333/pdf/C1333.pdf

Gurdak, J. J., & Roe, C. D. (2010). Recharge rates and chemistry beneath playas of the High Plains aquifer, USA. *Hydrogeology Journal, 18*, 1747–1772.

Guthery, F. S., & Bryant, F. C. (1982). Status of playas in the Southern Great Plains. *Wildlife Society Bulletin, 10*, 309–317.

Haaland, G. L., Matsushima, J. K., Johnson, D. E., & Ward, G. . (1981). Nutritive value of protected tallow product fed with hay to beef cattle as assess by respiration calorimetry. *J. Anim. Sci, 52*, 391–397.

Haan, M. M., Russell, J. R., Davis, J. D., & Morrical, D. G. (2010). Grazing management and microclimate effects on cattle distribution relative to a cool season pasture stream. *Rangeland Ecology & Management, 63*, 572–580.

Haas, H. J., Evans, C. E., & Miles, E. F. (1957). *Nitrogen and carbon changes in the Great Plains soils as influenced by cropping and soil treatments. Tech. Bull. No. 1164. Tech. Bull. No.* Washington, DC.

Hahn, G. L., Mader, T. L., & Gaughan, J. B. (1998). Proactive Feedlot Cattle Management During Heat Waves. *Anim. Prod. Aus., 22*, 318.

Hales, K. E., Cole, N. A., & MacDonald, J. C. (2012). Effects of corn processing method and dietary inclusion of wet distillers grains with solubles on energy metabolism, carbon-nitrogen balance, and enteric methane emissions of finishing cattle. *Journal of Animal Science, 90*(9), 3174–85.

Hall, T. L. (1998). The Success of Value Engineering on the Mni Wiconi Rural Water Project, South Dakota. In *Save International Conference Proceedings 1998*.

Halofsky, J. E., Peterson, D. L., O'Halloran, K. A., & Hawkins Hoffman, C. (2011). *Adapting to climate change at Olympic National Forest and Olympic National Park. Gen. Tech. Rep. PNW-GTR-844* (p. 130). Portland, OR.

Handmer, J., Dovers, S., & Downing, T. E. (1999). Societal vulnerability to climate change and variability. *Mitigation and Adaptation Strategies for Global Change, 4*(3-4), 267–281.

Hansen, A. J., Davis, C. R., Piekielek, N., Gross, J., Theobald, D., Goetz, S., … DeFries, R. (2011). Delineating the ecosystems containing protected areas for monitoring and management. *BioScience, 61*, 363–373.

Hansen, L., & Hoffman, J. (2010). *Climate Savvy: Adapting conservation and resource management to a changing world*. Washington, DC: Island Press.

Hanson, M. A., Zimmer, K. D., Butler, M. G., Tangen, B. A., Herwig, B. R., & Euliss, N. H. J. (2005). Biotic interactions as determinants of ecosystem structure in prairie wetlands: an example using fish. *Wetlands, 25*, 764–75.

Hartman, J. C., Nippert, J. B., Orozco, R. A., & Springer, C. J. (2011). Potential ecological impacts of switchgrass (Panicum virgatum L.) biofuel cultivation in the Central Great Plains, USA. *Biomass and Bioenergy, 35,* 3415–3421.

Hartman, M. D., Merchant, E. R., Parton, W. J., Gutmann, M. P., Lutz, S. M., & Williams, S. a. (2011). Impact of historical land-use changes on greenhouse gas exchange in the U.S. Great Plains, 1883-2003. *Ecological Applications : A Publication of the Ecological Society of America, 21*(4), 1105–19. Retrieved from http://www.ncbi.nlm.nih.gov/pubmed/21774417

Hartman, M. D., Merchant, E. R., Parton, W. J., Gutmann, M. P., Lutz, S. M., & Williams, S. A. (2011). Impact of Historical Land Use Changes in the U.S. Great Plains, 1883 to 2003. *Ecological Applications, 21*(4), 1105–1119.

Harvell, C. D., Altizer, S., Cattadori, I. M., Harrington, L., & Weil, E. (2009). Climate change and wildlife diseases: When does the host matter the most? *Ecology, 90,* 912–920.

Harvell, C. D., Mitchell, C. E., War, J. R., Altizer, S., Dobson, A. P., Ostfeld, R. S., & Samuel, M. D. (2002). Climate warming and disease risks for terrestrial and marine biota. *Science, 296,* 2158–2162.

Haslouer, S. G., Eberle, M. E., Edds, D. E., Gido, K. B., Mammoliti, C. S., Triplet, J. R., … Stark, W. J. (2005). Current status of native fish species in Kansas. *Transactions of the Kansas Academy of Science, 108,* 32–46.

Hassal, C., Thompson, D. J., G.C., F., & I.F., H. (2007). Historical changes in the phenology of British Odonata are related to climate. *Global Change Biology, 13,* 933–941.

Hassall, C., & Thompson, D. J. (2008). The effects of environmental warming on Odonata: a review. *International Journal of Odonatology, 11,* 131–153.

Haukos, D. A., & Smith, L. M. (1994). The importance of playa wetlands to biodiversity of the Southern High Plains. *Landscape and Urban Planning, 28,* 83–98.

Haukos, D. A., & Smith, L. M. (1997). *Common flora of the playa lakes* (p. 196). Lubbock, TX.

Havstad, K., Peters, D., Allen-Diaz, B., Bartolome, J., Bestelmeyer, B., Briske, D., … Yao, J. (2009). The Western United States rangelands: a major resource. In W. F. Wedin & S. L. Fales (Eds.), *Grasslands: quietness and strength for a New American Agriculture* (pp. 75–93). Madison, Wisconson, USA: American Society of Agronomy.

Hay, L. E., Markstrom, S. L., & Ward-Garrison, C. D. (2011). Watershed-scale response to climate change through the twenty-first century for selected basins across the United States. *Earth Interactions, 15,* 13.

Hegmon, M., Peeples, M. A., Kinzig, A. P., Kulow, S., M., M. C., & Nelson, M. C. (n.d.). Social transformation and its human costs in the prehispanic US Southwest. *2008, 110,* 313–324.

Heisler-White, J., Blair, J. M., Kelly, E., Harmoney, K., & Knapp, A. K. (2009). Contingent productivity responses to more extreme rainfall regimes across a grassland biome. *Global Change Biology, 15,* 2894–2904.

Helfman, G. S. (2007). *Fish Conservation: A Guide to Understanding and Restoring Global Aquatic Biodiversity and Fishery Resources*. Washington, DC, USA: Island Press.

Henderson, M. . S., & Robinson, D. L. (1982a). Environmental influences on fiber component concentrations of warm-season perennial grasses. *Agronomy Journal, 4,* 573–579.

Henderson, M. . S., & Robinson, D. L. (1982b). Environmental influences on yield and invitro trued igestibility of warm-season perennial grasses and the relationships to fiber components. *Agronomy Journal, 74,* 943–946.

Hendricks, P. (2003). Spring snow conditions, laying date, and clutch size in an alpine population of American Pipits. *J. Field Ornithol, 74*(423-429).

Herrero, M. P., & Johnson, R. R. (1980). High temperature stress and pollen viability in maize. *Crop Science, 20*(796-800).

Hickling, R., Roy, D. B., Hill, J. K., & Thomas, C. D. (2005). A northward shift of range margins in British Odonata. *Global Change Biology, 11*, 502–506.

Hobbs, R. J., Arico, S., & Aronson, J. et al. (2006). Novel ecosystems: theoretical and management aspects of the new ecological world order. *Global Ecol Biogeogr, 15*, 1–7.

Hobbs, R. J., & Huenneke, L. F. (1992). Disturbance, Diversity, and Invasion: Implications for Conservation. *Conservation Biology, 6*, 324–337.

Hoppe, R. A., & Banker, D. E. (2010). *Structure and Finances of U.S. Farms: Family Farm Report Economic Information Bulletin No. (EIB-66)* (p. 72). July.

Houser, S., Teller, V., MacCracken, M., Gough, R., & Spears, P. (2000). *Native Peoples-Native Homelands Climate Change Workshop Final Report. Oct 28- Nov 1, 1998* (pp. 351–377). Albuquerque, NM.

Houston City Council. (2011). *District G Newsletter. News from District G.* (p. 16). Aug-Sept 2011.

Hovenden, M. J., Newton, P. C. D., Carran, R. A., Theobald, P., Wills, K. E., Vander Schoor, J. K., … Osanai, Y. (2008). Warming prevents the elevated CO2-induced reduction in available soil nitrogen in a temperate, perennial grassland. *Global Change Biology, 14*, 1018–1024.

Hovenden, M. J., Wills, K. E., Vander Schoor, J. K., Williams, A. L., & Newton, P. C. D. (2008). Flowering phenology in a species-rich temperate grassland is sensitive to warming but not elevated CO2. *New Phytologist, 178*, 815–822.

Howell, T. A. (2009). Global climatic change effects on irrigation requirements for the Central Great Plains.In: Proceedings of the Central Plains Irrigation Conference and Exposition, February 24-25, 2009 (pp. 25–39). Colby, Kansas.

Hristov, A. N. (2012). Historic, pre-European settlement, and present-day contribution of wild ruminants to enteric methane emissions in the United States. *Journal of Animal Science, In Press.* doi:10.2527/jas.2011-4539

Hu, Q., Weiss, A., Feng, S., & Baenziger, P. S. (2005). Earlier winter wheat heading dates and warmer spring in the U.S. Great Plains. *Agricultural and Forest Meteorology, 135*, 284–290.

Huges, R. M., Rinne, J. N., & Calamusso, B. (2005). Introduction to historical changes in large river fish assemblages of the Americas. In J. N. Rinne, R. M. Hughes, & B. Calmusso (Eds.), *Historical changes in large river fish assemblages of the Americas* (pp. 1–12). Bethesda, MD: American Fisheries Society, Symposium 45.

Huntjens, P., Lebel, L., Pahl-Wostl, C., Camkin, J., Schulze, R., & Kranz, N. (2012). Institutional design propositions for the governance of adaptation to climate change in the water sector. *Global Environmental Change, 22*(1), 67–81. doi:10.1016/j.gloenvcha.2011.09.015

Huntjens, P., Pahl-Wostl, C., & Grin, J. (2010). Climate change adaptation in European river basins. *Regional Environmental Change, 10*(4), 263–284.

Huntzinger, T. L. (1995). Surface Water: A Critical Resource of the Great Plains. In S. R. Johnson & A. Bouzaher (Eds.), *Conservation of Great Plains Ecosystems* (pp. 253–274). Dordrecht, Netherlands: Kluwer Academic Publishers.

Huxman, T. E., Smith, M. D., Fay, P. A., Knapp, A. K., Shaw, M. R., Loik, M. E., … Williams, D. G. (2004). Convergence across biomes to a common rain-use efficiency. *Nature, 429*, 651–654.

International Board for Soil Research and Management. (1990). Organic-matter management and tillage in humid and sub-humid Africa. In *IBSRAM Proceedings No. 10*. Bangkok, Thailand.

International Institute for Sustainable Development, International Union for Conservation of Nature and Natural Resources, & Stockholm Environment Institute. (2003). *Livelihoods and Climate Change, Combining Disaster Risk Reduction, Natural Resource Management and Climate*

Change Adaptation in a New Approach to the Reduction of Vulnerability and Poverty. Manitoba, Canada.

Intertribal Climate Change Working Group. (2009). *A Tribal White Paper on Climate Change Adaptation and Mitigation.*

Intertribal Council on Utility Policy. (n.d.). *Energy Justice in Native America: A Policy Paper for Consideration by the Obama Administration and the 111th Congress.* Retrieved from http://www.treatycouncil.org/PDF/EJ in NA Policy Paper.pdf

IPCC. (2007). GENERAL GUIDELINES ON THE USE OF SCENARIO DATA FOR CLIMATE IMPACT AND. *Environment*, (June).

IPCC. (2012). *Managing the Risks of Extreme Events and Disasters to Advance Climate Change Adaptation (SREX).*

Ison, R., Roling, N., & Watson, D. (2007). Challenges to science and society in the sustainable management and use of water: investigating the role of social learning. *Environmental Science and Policy*, *10*(6), 499–511.

Jelks, H. L., Walsh, S. J., Burkhead, N. M., Conteras-Balderas, S., Diaz-Pardo, E., Hendrickson, D. A., … Warren, M. L. J. (2008). Conservation status of imperiled North American freshwater and diadromous fishes. *Fisheries*, *33*, 372–407.

Johnson, B. L. (1999). The role of adaptive management as an operational approach for resource management agencies. *Conservation Ecology*, *3*(2), 8.

Johnson, J. M. F., Archer, D. W., Weyers, S. L., & Barbour, N. W. (2011). Do mitigation strategies reduce global warming potential in the Northern U.S. Corn Belt? *Journal of Environmental Quality*, *40*, 1551–1559.

Johnson, P. T. J., & Thieltges, D. W. (2010). Diversity, decoys and the dilution effect: How ecological communities affect disease risk. *J. Exp. Biol.*, *213*, 961–970.

Jones, K. E., Patel, N. G., Levy, M. A., Storeygard, A., Balk, D., Gittleman, J. L., & Daszak, P. (2008). Global trends in emerging infectious diseases. *Nature*, *451*(7181), 990–993.

Jones, R. N., & Preston, B. L. (2011). Adaptation and risk management. *Wiley Interdisciplinary Reviews: Climate Change*, *2*(2), 296–308.

Joyce, L. A., Blate, G. M., Littell, J. S., McNulty, S. G., Millar, C. I., Moser, S. C., … Peterson, D. L. (2009). Managing for multiple resources under climate change. *Environmental Management*, *44*, 1022–1032.

Julius, S., West, J., Baron, J., Joyce, L., Kareiva, P., Keller, B., … Scott, J. M. (2008). *Preliminary review of adaptation options for climate-sensitive ecosystems and resources. Synthesis and assessment product 4.4 by the U.S. climate change science program and the subcommittee on global change research* (p. 873). Washington, DC, USA.

Kallis, G. (2008). Droughts. *Annual Review of Environment and Resources*, *33*, 85–118.

Kaplan, S. M. (2009). *Electric Power Transmission: Background and Policy Issues*. Retrieved from http://www.policyarchive.org/handle/10207/bitstreams/18743.pdf

Karl, T. R., Melillo, J. M., & Peterson, T. C. (eds). (2009). *Global Climate Change Impacts in the United States.*

Karl, T. R., & Quayle, R. G. (1981). The 1980 summer heat wave and drought in historical perspective. *Mon. Wea. Rev.*, *109*, 2055–2073.

Kelly, P. M., & Adger, W. N. (2000). Theory and practice in assessing vulnerability to climate change and facilitating adaptation. *Climatic Change*, *47*(4), 325–352.

Kenney, D. S., Goemans, C., Klein, R., Lowrey, J., & Reidy, K. (2008). Residential water demand management: Lessons from Aurora, Colorado. *Journal of the American Water Resources Association*, *44*(1), 192–207.

Kenny, J. F., Barber, N. L., Hutson, S. S., Linsey, K. S., Lovelace, J. K., & Maupin, M. A. (2009). *Estimated Use of Water in the United States in 2005: U.S. Geological Survey Circular 1344* (p. 52).

Kimball. (1983). Carbon dioxide and agricultural yield: an assemblage and analysis of 430 prior observations. *Agronomy Journal, 75*, 779–788.

King, J. Y., Mosier, A. R., Morgan, J. A., LeCain, D. R., Milchunas, D. G., & Parton, W. J. (2004). Plant nitrogen dynamics in short grass steppe under elevated atmospheric carbon dioxide. *Ecosystem, 7*, 147–160.

Kittel, T. G. F., Rosenbloom, N. A., Painter, T. H., Schimel, D. S., & VEMAP Modeling Participants. (1995). The VEMAP integrated database for modeling United States ecosystem /vegetation sensitivity to climate change. *J. Biogeography, 22*, 857–862.

Knapp, A. K., Beier, C., Briske, D. D., Classen, A. T., Luo, Y., Reichstein, M., ... Weng, E. (2008). Consequences of more extreme precipitation regimes for terrestrial ecosystems. *BioScience, 58*, 811–821.

Knapp, A. K., & Smith, M. D. (2001). Variation among Biomes in Temporal Dynamics of Above-ground Primary Production. *Science, 291*, 481–484.

Knight, D. B., & David, R. E. (2009). Contribution of tropical cyclones to extreme rainfall events in the southeastern United States. *J Geophys Res, 114*, D23102.

Knopf, F. L., & Smson, F. B. (1997). Conservation of grassland vertebrates. *Ecological Studies, 125*, 273–289.

Knutson, C. (2008). Adaptation to water stress in the Great Plains. *Water Resources IMPACT, 10*(4), 18–21.

Knutson, C. L., & Haigh, T. et al. (2011). Farmer perceptions of sustainable agriculture practices and drought risk reduction in Nebraska, USA. *Renewable Agriculture and Food Systems, 26*(3), 255–266.

Knutson, C. L., Jones, S., Sittler, M. E., Higgins, M., Svoboda, M. D., Kluck, D. R., & Wilhite, D. A. (2008). Development of a Low Flow Early Warning System for the NOAA National Weather Service, USA, Stream Ecology and Low Flows. *Special Issue of the International Journal of Ecological Economics and Statistics, 12*(F08), 44–57.

Koch, B., Khosla, M., Westfall, M. D., Inman, D., & Frasier, W. M. (2004). Economic Feasibility of Variable-Rate Nitrogen Application in Site-Specific Management Zones. *Agronomy Journal, 96*, 1572–1580.

Koons, W. (1992). Amphibians in Manitoba. *Canadian Wildlife Service Occasional Paper, 76*, 19–20.

Koshmrl, M. (2011). Harnessing the Sun, Lakota Style. *Solar Today*, March.

Kostyack, J., Lawler, J. J., Goble, D. D., Olden, J. D., & Scott, J. M. (2011). Beyond Reserves and Corridors: Policy Solutions to Facilitate the Movement of Plants and Animals in a Changing Climate. *BioScience, 61*, 716–719.

Kovats, R. S., & Hajat, S. (2008). Heat Stress and Public Health: A Critical Review. *Annual Review of Public Health.*

Kruk, M. C., Gibney, E. J., Levinson, D. H., & Squires, M. (2010). A Climatology of inland winds from tropical cyclones for the eastern United States. *J. Appl. Meteor. Climatol., 49*, 1538–1547.

Kundzewicz, Z. W., Mata, L. J., Arnell, N. W., Doll, P., Jimenez, B., Miller, K., ... Shiklomanov, I. (2008). The implications of projected climate change for freshwater resources and their management. *Hydrol. Sci. J., 53*, 3–10.

Kunkel, K. E., Easterling, D. R., Kristovich, A. R., Gleason, B., Stoecker, L., & Smith, R. (2010). Recent increases in US heavy precipitation associated with tropical cyclones. *Geophys. Res. Lett., 37*, L24706.

Kunkel, K. E., Karl, T. R., Brooks, H., Kossin, J., Lawrimore, J. H., Arndt, D., … Wuebbles, D. (2013). Monitoring and understanding trends in extreme storms: State of knowledge. *Bulletin of the American Meteorological Society, 94*, 499–514.

Lackett, J. M., & Galvin, K. A. (2008). From Fragmentation to Reaggregation of Rangelands in the Northern Great Plains, USA. In K. A. Galvin, R. S. Reid, R. H. Behnke, & N. T. Hobbs (Eds.), *Fragmentation of Semi-Arid and Arid Landscapes. Consequences for Human and Natural Systems* (pp. 113–134). Corthecht, The Netherlands: Springer.

Lafferty, K. D. (2009). The ecology of climate change and infectious diseases. *Ecology, 90*, 888–900.

Lafferty, K. D. (2010). Interacting parasites. *Science, 330*, 18–188.

LaGrange, T. G., Stutheit, R., Gilbert, M., Shurtliff, D., & Whited, P. M. (2011). *Sedimentation of Nebraska's Wetlands: Review of Current Knowledge and Issues* (p. 62). Lincoln, Nebraska.

Lal, P., Alavalapati, J., & Mercer, D. E. (2011). Socioeconomic Impacts of Climate Change on Rural Communities in the United States in Effects of Climate Change on Natural Resources and Communities. In *A Compendium of Briefing Papers*.

Lal, R. (1979). Importance of tillage systems in soil and water management in the tropics. In: Soil Tillage and Crop Production. In R. Lal (Ed.), *IITA Proc. Ser. 2* (pp. 25–32).

Lal, R. (2001). Potential of desertification control to sequester carbon and mitigate the greenhouse effect. *Climatic Change, 51*, 35–72.

Landesman, W. J., Allan, B. F., Langerhans, R. B., Knight, T. M., & Chase, J. M. (2007). Inter-annual associations between precipitation and human incidence of West Nile virus in the United States. *Vector-Borne and Zoonotic Diseases, 7*, 337–343.

Lauenroth, W. K., & Burke, I. C. (1995). The Great Plains, climate variability. In *Encyclopedia of environmental biology* (2nd ed., pp. 237–249). San Diego, CA: Academic Press.

Lauenroth, W. K., Burke, I. C., & Paruelo, J. M. (2000). Patterns of production and precipitation-use efficiency of winter wheat and native grasslands in the Central Great Plains of the United States. *Ecosystems, 3*, 344–351.

Lauenroth, W. K., & Milchunas, D. G. (1991). Short-grass steppe. In R. T. Coupland (Ed.), *Natural Grasslands: Introductions and western hemisphere* (pp. 183–226). New York, NY: Elsevier.

Lawrence, D. J., Larson, E. R., Liermann, C. A. R., Mims, M. C., Pool, T. K., & Olden, J. D. (2011). National parks as protected areas for U.S. freshwater fish diversity. *Conservation Letters, 4*, 364–371.

Leach, M. K., & Givnish, T. . . (1996). Ecological determinants of species loss in remnant prairies. *Science, 273*, 1555–1558.

Leakey, A. D. B. (2009). Rising atmospheric carbon dioxide concentration and the future of C4 crops for food and fuel. *Society, Proceedings of the Royal, B*(276), 2333–2343.

Lebel, L., Grothmann, T., & Siebenhüner, B. (2010). The role of social learning in adaptiveness: insights from water management. *International Environmental Agreements: Politics, Law and Economics, 10*(4), 333–353.

Lee, E. T., Howard, B. V., Savage, P. J., Cowan, L. D., Fabsitz, R. R., Oopik, A. J., … Welty, T. K. (1995). Diabetes and impaired glucose tolerance in three American Indian populations aged 45-74 years. *Diabetes Care, 18*(5), 599–610.

Lehmann, E. (2011). Violent Storms Makes 2011 "One for the Record Books," Insurance Price Hikes Under Way. *Climate Wire from E&E Publishing*, p. 13 July.

Lett, M. S., & Knapp, A. K. (2005). Woody plant encroachment and removal in mesic grassland: Production and composition responses of herbaceous vegetation. *American Midland Naturalist, 153*, 217–231.

Lett, M. S., Knapp, A. K., Briggs, J. M., & Blair, J. M. (2004). Influence of shrub encroachment on aboveground net primary productivity and carbon and nitrogen pools in a mesic grassland. *Canadian Journal of Botony, 82*, 1363–1370.

Lettenmaier, D., Major, D., Poff, L., & Running, S. (2008). *Water Resources: The Effects of Climate Change on Agriculture, Land Resources, Water Resources, and Biodiversity in the United States* (p. 362). Washington, DC, USA.

Liebig, M. A., Gross, J. R., Kronberg, S. L., Phillips, R. L., & Hanson, J. D. (2010). Grazing management contributions to net global warming potential: a long-term evaluation in the Northern Great Plains. *Journal of Environmental Quality, 39*, 799–809.

Lorenz, D. (2010). The diversity of resilience: contributions from a social science perspective. *Natural Hazards*, 1–18.

Low, S. A., & Vogel, S. (2011). *Direct and Intermediated Marketing of Local Food in the United States, ERR-128* (p. November).

Luers, A. L. (2005). The surface of vulnerability: an analytical framework for examining environmental change. *Global Environmental Change, 15*, 214–223.

Luo, H. R., Smith, L. M., Allen, B. L., & Haukos, D. A. (1997). Effects of sedimentation on playa wetland volume. *Ecological Applications, 7*, 247–252.

Luo, Y., Sherry, R., Zhou, X., & Wan, S. (2009). Terrestrial carbon-cycle feedback to climate warming: experimental evidence on plant regulation and impacts of biofuel feedstock harvest. *GCB Bioenergy, 1*, 62–74.

Luo, Z. B., Li, K., Jiang, X., & Polle, A. (2009). Ectomycorrhizal fungus (Paxillus involutus) and hydrogels affect performance of Populus euphratica exposed to drought stress. *Ann. For. Sci., 66*(1), 106.

Lynch, J. D. (1978). The distribution of leopard frogs (Rana blairi and Rana pipiens) (Amphibia, Anura, Ranidae) in Nebraska. *Journal of Herpetology, 12*, 157–162.

Lytle, D. A., & Poff, N. L. (2004). Adaptation to natural flow regimes. *Trends in Ecology and Evolution, 19*, 94–100.

Lyytimäki, J., & Hildén, M. (2007). Thresholds of sustainability: policy challenges of regime shifts in coastal areas. Sustainability: Science, Practice & Policy 3. In *http://ejournal.nbii.org* (pp. 61–69).

M.J., L., Lang, K., Waltz, T., & Phillips, G. S. (1994). An altered amphibian assemblage: Dickenson County, Iowa, seventy years after Frank Blanchard's survey. *American Midland Naturalist, 131*, 311–319.

MacKnick, J., Newmark, R., Heath, G., & Hallet, K. C. (2011). *A review of operational water consumption and withdrawal factors for electricity generating technologies*. Golden, CO.

Mackun, P., & Wilson, S. (2011). *Population Distribution and Change: 2000 to 2010. Census Brief C2010BR-01* (p. March). Retrieved from http://www.census.gov/prod/cen2010/briefs/c2010br-01.pdf

Mader, T. L. (2003). Environmental stress in confined beef cattle. *J. Anim. Sci, 81*, E110–E119.

Malewitz, J. (2011, October 7). Persistent drought threatens Texas oil industry. *Stateline*. Retrieved from http://www.stateline.org/live/details/story?contentId=605095

Marchetti, M. P., & Moyle, P. B. (2001). Effects of flow regime and habitat structure on fish assemblages in a regulated California stream. *Ecological Applications, 11*, 530–539.

Markstrom, S. L., Hay, L. E., Ward-Garrison, C. D., Risley, J. C., Battaglin, W. A., Bjerklie, D. M., ... Walker, J. F. (2011). *An integrated watershed scale response to climate change for selected basins across the United States: U.S. Geological Survey Scientific Investigations Report 2011–5077* (p. 142).

Markstrom, S. L., Niswonger, R. G., Regan, R. S., Prudic, D. E., & Barlow, P. M. (2008). *GSFLOW: Coupled ground-water and surface-water flow model based on the integration of the Precipitation-Runoff Modeling System (PRMS) and the Modular Ground-Water Flow Model (MODFLOW-2005).* (pp. 6–D1).

Martens, W. J. M., Jetten, T. H., & Focks, D. A. (1997). Sensitivity of malaria, schistosomiasis and dengue to global warming. *Climatic Change, 35,* 145–156.

Martin, C., Morgan, D. P., & Doreau, M. (2010). Methane mitigation in ruminants from microbe to the farm scale. *Animal, 4,* 351–365.

Martin, P. H., & Lefebvre, M. C. (1995). Malaria and climate: sensitivity of malaria potential transmission to climate. *Anihio, 24,* 200–209.

Matthews, W. J. (1988). North American prairie streams as systems for ecological study. *Journal of North American Benthol Soc, 7,* 387–409.

Maupin, M. A., & Barber, N. L. (2005). *Estimated withdrawals from principal aquifers in the United States, 2000: U.S. Geological Survey Circular 1279* (p. 52).

Mawdsley, J. (2011). Design of conservation strategies for climate adaptation. *WIREs Clim Change, 2,* 498–515.

Maxwell, R. M., & Kollet, S. J. (2008). Interdependence of groundwater dynamics and land-energy feedbacks under climate change. *Nature Geoscience, 1,* 665–669.

May, B., & Plummer, R. (2011). Accommodating the challenges of climate change adaptation and governance in conventional risk management: adaptive collaborative risk management (ACRM). *Ecology and Society, 16*(1), 47.

Maynard, N. G. (1998). Native Peoples-Native Homelands. In *Climate Change Workshop. Circles of Wisdom.*

McBride, W. D., & Matthews, K. (2011). *The diverse structure and organization of U.S. beef cow-calf farms. EIB No. 73.* Washington, DC. Retrieved from http://www.ers.usda.gov/publications/eib73/

McCabe, G. J., Ault, T. R., Cook, J., Betancourt, J. L., & Schwartz, M. D. (2011). Influences of the El Nino Southern Oscillation and the Pacific Decadal Oscillation on the timing of the North American spring. *International Journal of Climatology, 32,* 2301–2310.

McCauley, S. J. (2006). The effects of dispersal and recruitment limitation on community structure of Odonates in artificial ponds. *Ecography, 29,* 585–595.

McDonald, R. I., Fargione, J., Kiesecker, J., Miller, W. M., & Powell, J. (2009). Energy Sprawl or Energy Efficiency: Climate Policy Impacts on Natural Habitat for the United States of America. *PloS One, 4*(8), e6802.

McGeehin, M. A., & Mirabelli, M. (2001). The potential impacts of climate variability and change on temperature-related morbidity and mortality in the United States. *Environ. Health Persp., 109,* 185.

McGuire. (2007). *Water-level Changes in the High Plains Aquifer, Predevelopment to 2005 and 2003 to 2005. U.S. Geological Survey scientific investigations report 2006–5324* (p. 7). Reston, VA. Retrieved from http://pubs.usgs.gov/sir/2006/5324/>

McGuire, V. (2011). *Water-level changes in the High Plains aquifer, predevelopment to 2009, 2007-08, and 2008-09, and change in water storage, predevelopment to 2009: U.S. Geological Survey Scientific Investigations Report 2011-5089: 13.*

McLauchlan, M. (2008). McLauchlan, M. Retrieved from http://www.fws.gov/filedownloads/ftp_Region6_upload/Megan McLachlan/PLJV_Maps_Data/Probable_Playas_v2

McLaughlin, S. (2010). Tradition and Diabetes Prevention: A Healthy Path for Native Americans. *Diabetes Spectrum – Special Report, 23*(4).

McLean, R. G. (2008). The introduction and emergence of wildlife diseases in North America. In T. E. Fulbright & D. G. Hewitt (Eds.), *Wildlife science: Linking ecological theory and management implications* (pp. 261–278). Boca Raton, Florida: CRC Press.

McLeman, R., Mayo, D., Strebeck, E., & Smit, B. (2008). Drought adaptation in rural eastern Oklahoma in the 1930s: Lessons for climate change adaptation research. *Mitigation and Adaptation Strategies for Global Change, 13*(4), 379–400.

McLeman, R., & Smit, B. (2006). Migration as an adaptation to climate change. *Climatic Change, 76*, 31–53.

McMahon, P. B., Dennehy, K. F., Bruce, B. W., Gurdak, J. J., & Qi, S. L. (2007). *Water-quality assessment of the High Plains Aquifer, 1999–2004: U.S. Geological Survey Professional Paper 1749* (p. 136). Retrieved from http://pubs.usgs.gov/pp/1749/

McNeeley, S. (2011). Examining barriers and opportunities for sustainable adaptation to climate change in Interior Alaska. *Climatic Change, 111*, 835–857.

McNeeley, S., & Shulski, M. (2011). Anatomy of a closing window: Vulnerability to changing seasonality in Interior Alaska. *Global Environmental Change, 21*(2), 464–473.

Mehta, V. M., Rosenberg, N. J., & Mendoza, K. (2011). Simulated Impacts of Three Decadal Climate Variability Phenomena on Water Yields in the Missouri River Basin. *Journal Of The American Water Resources Association, 47*, 126–135.

Mid-America Regional Council. (n.d.). *Best Practices for Stormwater Management in Metropolitan Kansas City – Quinton Heights Neighborhood Redevelopment*. Topeka, KS.

Milchunas, D. G., Mosier, A. R., Morgan, J. A., Lecain, D. R., King, J. Y., & Nelson, J. A. (2005). Elevated CO2 and defoliation effects on a shortgrass steppe: forage quality versus quantity for ruminants. *Agriculture, Ecosystems and Environment, 111*, 166–184.

Miller, M. E. (2005). *The structure and functioning of dryland ecosystems - conceptual models to inform long-term ecological monitoring.* (pp. 2005–5197).

Mills, E. (2009). A Global Review of Insurance Industry Responses. *The Geneva Papers, 34*, 323–359.

Mills, E., Roth Jr., R. J., & Lecomte, E. (2006). Availability and affordability of insurance under climate change: A growing challenge for the U.S. *Journal of Insurance Regulation, 25*(2), 109–149.

Milly, P. C. D., Betancourt, J., Falkenmark, M., Hirsch, R., Kundzewicz, Z. W., Lettenmaier, D. P., & Stouffer, R. J. (2008). Stationarity is Dead: Wither Water Management? *Science, 319*(5863), 573–574.

Mishra, G. S., & Yeh, S. (2011). Life Cycle Water Consumption and Withdrawal Requirements of Ethanol from Corn Grain and Residues. *Environ. Sci. Technol., 45*(10), 4563–4569.

Morgan, J. A., Derner, J. D., Milchunas, D. G., & Pendall, E. (2008). Management implications of global change for Great Plains rangelands. *Rangelands, 30*(3), 18–22.

Morgan, J. A., LeCain, D. R., Pendall, E., Blumenthal, D. M., Kimball, B. A., Carillo, Y., … Dijkstra, F. (2011). Elevated CO2 eliminates desiccating effects of simulated warming in semi-arid grassland. *Nature, 476*, 202–205.

Morgan, J. A., LeCain, D. R., Pendall, E., Blumenthal, D. M., Kimball, B. A., & Carrillo, Y., et al. (2011). C4 grasses prosper as carbon dioxide eliminates desiccation in warmed semi-arid grassland. *Nature, 476*, 202–205.

Morgan, J. A., Milchunas, D. G., Lecain, D. R., West, M., & Mosier, A. R. (2007). Carbon dioxide enrichment alters plant community structure and accelerates shrub growth in the shortgrass steppe. *Proceedings of the National Academy of Sciences, 104*, 14724–14729.

Morgan, J. A., Mosier, A. R., Milchunas, D. G., Lecain, D. R., Nelson, J. A., & Parton, W. J. (2004). CO2 enhances productivity, alters species composition, and reduces digestibility of shortgrass steppe vegetation. *Ecological Applications, 14*, 208–219.

Morgan, J. A., Pataki, D. E., Körner, C., Clark, H., Del Grosso, S. J., Grünzweig, J. M., ... Shaw, M. R. (2004). Water relations in grassland and desert ecosystems exposed to elevated atmospheric CO2. *Oecologia, 140*, 11–25.

Mosier, A. R., Delgado, J. A., Cochran, V. L., Valentine, D. W., & Parton, W. J. (1997). Impact of agriculture on soil consumption of atmospheric CH4 and a comparison of CH4 and N2O flux in subarctic, temperate and tropical grasslands. *Nutr. Cycl. Agroecosyst., 49*, 71–83.

Mosier, A. R., Morgan, J. A., King, J. Y., LeCain, D., & Milchunas, D. G. (2002). Soil–atmosphere exchange of CH4, CO2, NOx, and N2O in the Colorado shortgrass steppe under elevated CO2. *Plant and Soil, 240*, 201–211.

Mosier, A. R., Pendall, E., & Morgan, J. A. (2003). Soil-atmosphere exchange of CH4, CO2, NOx, and N2O in the Colorado Shortgrass Steppe following five years of elevated CO2 and N fertilization. *Atmos. Chem. Phys. Discuss., 3*, 2691–2706.

Moss, A. R., Jouany, J. P., & Newbold, J. (2000). Review Article: Methane production by ruminants: its contribution to global warming. *Ann. Zootech, 49*, 231–253.

Muller, D., & Smith, R. (2000). *Arkansas River Water Needs Assessment. Section 2 – Introduction* (p. July).

Nagler, A., Bastian, C., Hewlett, J. P., Mooney, S., Paisley, S. I., Smith, M. A., ... Ponnamenen, P. (2007). *Multiple Impacts, Multiple Strategies: How Wyoming Cattle Producers are Surviving in Prolonged Drought*. Laramie, WY.

Nardone, A., Ronchi, B., Lacetera, N., Ranieri, M. S., & Bernabucci, U. (2010). Effects of climate change on animal production and sustainability of livestock systems. *Livestock Science, 130*, 57–69.

National Association of Mutual Insurance Companies. (2011). Insurer Initiatives: Insurance Company Climate Change Initiatives.

National Climate Assessment Synthesis Team. (2001). *Climate Change Impacts on the United States: The Potential Consequences of Climate Variability and Change* (p. 620). Cambridge, U.K.

National Drought Mitigation Center. (2010). *Republican River Basin Water and Drought Portal.*

National Drought Mitigation Center. (2012). *Managing Drought Risk on the Ranch. 2012.*

National Oceanic and Atmospheric Administration. (2011a). *Extreme Weather 2011.*

National Oceanic and Atmospheric Administration. (2011b). *State of the Climate: Drought* (p. August).

National Oceanic and Atmospheric Administration. (2011c). *State of the Climate: National Overview for August 2011* (p. September). Retrieved from http://www.ncdc.noaa.gov/sotc/2011/8

National Oceanic and Atmospheric Administration. (2012). Billion Dollar U.S. Weather / Climate Disasters. Retrieved from http://www.ncdc.noaa.gov/oa/reports/billionz.html#chron

National Park Service. (2010). *National Park Service Climate Change Response Strategy*. Fort Collins, CO.

National Renewable Energy Laboratory. (2009). Greensburg Kansas: A Better Place to Live.

National Renewable Energy Laboratory. (2012a). Dynamic Maps, GIS Data, & Analysis Tools: Geothermal Maps.

National Renewable Energy Laboratory. (2012b). Dynamic Maps, GIS Data, & Analysis Tools: Solar Maps.

National Renewable Energy Laboratory. (2012c). Dynamic Maps, GIS Data, & Analysis Tools: Wind Maps.

National Research Council. (2002). *The Missouri River Ecosystem: Exploring the Prospects for Recovery*. Washington, DC, USA.

National Research Council. (2008). *Desalination: A National Perspective*. Washington, DC, USA: National Academies Press.

National Research Council. (2011). *Water Reuse: Expanding the Nation's Water Supply Through Reuse of Municipal Wastewater*. Washington, DC, USA: National Academies Press.

Nelson, R. W., Logan, W. J., & Weller, E. C. (1983). *Playa wetlands and wildlife on the Southern Great Plains: A characterization of habitat. OBS 83/28*. Albuquerque, NM.

Newman, Y. C., Sinclair, T. R., Blount, A. S., Lugo, M. L., & Valencia, E. (2007). Forage Production of Tropical Grasses under Extended Daylength at Subtropical and Tropical Latitudes. *Environmental and Experimental Botony, 61*(1), 18–24.

Nichols, J. D., Koneff, M. D., Heglund, P. J., Knutson, M. G., Seamans, M. E., Lyons, J. E., ... Williams, B. K. (2011). Climate change, uncertainty, and natural resource management. *Journal of Wildlife Management, 75*, 6–18.

NIDIS. (2011). National Integrated Drought Information System. Drought Portal. Retrieved January 03, 2012, from www.drought.gov/

Nielsen-Gammon, J. W. (2011). *The 2011 Texas Drought: A Briefing Packet for the Texas Legislature*. College Station, TX.

Niklaus, P. A., Kandele, r E., Leadley, P. W., Schmid, B., Tscherko, D., & Ko¨rner, C. (2004). A link between plant diversity, elevated CO2 and soil nitrate. *Oecologia, 127*, 540–548.

NOAA. (2011). State of the Climate Drought August 2011. Retrieved from http://www.ncdc.noaa.gov/sotc/drought/2011/8

Norgaard, R. B., Kallis, G., & Kiparsky, M. (2009). Collectively engaging complex socio-ecological systems: re-envisioning science, governance, and the California Delta. *Environmental Science and Policy, 12*(6), 644–652.

Norris, M. D., Blair, J. M., & Johnson, L. C. (2001). Land cover change in eastern Kansas: litter dynamics of closed-canopy eastern redceder forests in tallgrass prairie. *Canadian Journal of Botony, 79*, 214–222.

Norris, M. D., Blair, J. M., Johnson, L. C., & McKane, R. B. (2001). Assessing changes in biomass, productivity, and C and N stores following Juniperus virginiana forest expansion into tallgrass prairie. *Canadian Journal of Forest Research, 31*, 1940–1946.

North American Bird Conservation Initiative - U.S. Committee. (2010). *The State of the Birds 2010 Report on Climate Change, United States of America* (p. 32). Washington, DC, USA.

North American Bird Conservation Initiative - U.S. Committee. (2011). *The State of the Birds 2011 Report on Public Lands and Waters* (p. 48). Washington, DC, USA.

North Dakota Association of Rural Electric Cooperatives. (2012). North Dakota Storm Report. Retrieved from http://www.ndarec.com/mediagallery/galleryaprilstorm.html

Northern Cheyenne Tribe. (2007). *Northern Cheyenne Tribe Drought Mitigation Plan* (p. 34). Retrieved from http://cheyennenation.com/executive/drought_plan_june2007.pdf

Northern Water. (2012a). Colorado-Big Thompson History. Retrieved from http://www.northernwater.org/AboutUs/C-BTHistory.aspx

Northern Water. (2012b). Windy Gap Project History. Retrieved from http://www.northernwater.org/AboutUs/WindyGapHistory.aspx

Norwood, C. A. (2000). Water Use and Yield of Limited-Irrigated and Dryland Corn. *Soil Science Society of America Journal, 64*, 365–370.

Nutter, F. (2011). *21st century approach needed for NFIP reform* (p. 30 June).

O'Brien, K., Leichenko, R., Kelkar, U., Venema, H., Aandahl, G., Tompkins, H., … West, J. (2004). Mapping vulnerability to multiple stressors: climate change and globalization in India. *Global Environmental Change*, *14*(4), 303–313. doi:10.1016/j.gloenvcha.2004.01.001

O'Driscoll, P., & Kenworthy, T. (2005, April 28). Western drought shrinking Big Muddy. *USA Today*.

Oertli, B. (2008). The use of dragonflies in the assessment and monitoring of aquatic habitat. In A. Córdoba-Aguilar (Ed.), *Dragonflies & Damselflies: Model Organisms for Ecological and Evolutionary Research* (pp. 79–95). New York, NY: Oxford University Press.

Ogden, N. H., Lindsay, L. R., Hanincova, K., Barker, I. K., Bigras-Poulin, M., Charron, D. F., … Thompson, R. A. (2008). Role of migratory birds in introduction and range expansion of Ixodes scapularis ticks and of Borrelia burgdorferi and Anaplasma phagocytophilum in Canada. *Appl. Environ. Microbiol.*, *74*, 1780–1790.

Ojima, D., Garcia, L., Elgaali, E., Miller, K., Kittel, T. G. F., & Lackett, J. (1999). Potential Climate Change Impacts on Water Resources in the Great Plains. *Journal Of The American Water Resources Association*, *35*(6), 1443–1454. doi:10.1111/j.1752-1688.1999.tb04228.x

Ojima, D. S., & Corell, R. W. (2009). Managing Grassland Ecosystems under Global Environmental Change: Developing Strategies to Meet Challenges and Opportunities of Global Change. In A. J. Franzluebbers (Ed.), *Farming with Grass: Achieving Sustainable Mixed Agricultural Landscapes* (pp. 146–15). Ankeny, IA: Soil and Water Conservation Society.

Ojima, D. S., & Lackett, J. M. (2002). *Preparing for a changing climate: The Potential Consequences of Climate Variability and Change. Change.*

Ojima, D. S., Valentine, D. W., Mosier, A. R., Parton, W. J., & Schimel, D. S. (1993). Effect of land use change on methane oxidation in temperate forest and grasslands. *Chemosphere*, *26*, 675–685.

Okayasu, T., Okuro, T., Jamsran, U., & Takeuchi, K. (2011). Threshold distinctions between equilibrium and nonequilibrium pastoral systems along a continuous climatic gradient. *Rangeland Ecology & Management*, *64*, 10–17.

Osbahr, H., Twyman, C., Adger, W. N., & Thomas, D. S. G. (2010). Evaluating Successful Livelihood Adaptation to Climate Variability and Change in Southern Africa. *Ecol. Soc.*, *15*(2).

Ostlie, W. R., Schneider, R. E., Aldrich, J. M., Faust, T. M., McKim, R. L. B., & Chaplin, J. S. (1997). *The status of biodiversity in the Great Plains*. Minneapolis, Minnesota, USA.

Owensby, C. E., Ham, J. M., Knapp, A. K., & Auen, L. M. (1999). Biomass production and species composition change in a tallgrass prairie ecosystem after long-term exposure to elevated atmospheric CO2. *Global Change Biology*, *5*, 497–506.

Pahl-Wostl, C. (2007). The implications of complexity for integrated resources management. *Environmental Modelling & Software*, *22*(5), 561–569. doi:10.1016/j.envsoft.2005.12.024

Park, T., Ahearn, M., Covey, T., Erickson, K., Harris, J. M., Ifft, J., … Williams, R. (2011). *Agricultural Income and Finance Outlook / AIS-91* (p. December 2011).

Parmesan, C. (2006). Ecological and Evolutionary Responses to Recent Climate Change. *Annual Review of Ecology, Evolution, and Systematics*, *37*(1), 637–669.

Parmesan, C. (2007). Influences of species, latitudes and methodologies on estimates of phenological response to global warming. *Global Change Biology*, *13*, 1860–1872.

Parmesan, C., & Yohe, G. (2003). A globally coherent fingerprint of climate change impacts across natural systems. *Nature*, *421*(6918), 37–42. doi:10.1038/nature01286

Parton, W. J., Gutmann, N. P., Williams, S. A., Easter, M., & Ojima, D. (2005). Ecological impact of historical land-use patterns in the Great Plains: a methodological assessment. *Ecological Applications*, *15*(6), 1915–1928.

Parton, W. J., Myron, P., & Ojima, D. (2007). Long-term Trends in Population , Farm Income , and Crop Production in the Great Plains. *BioScience*, *57*(9), 737–747.

Parton, W. J., Myron, P., & Ojima, D. (2007). Long-term Trends in Population, Farm Income, and Crop Production in the Great Plains. *BioScience*, *57*(9), 737–747.

Parton, W. J., Ojima, D. S., & Schimel, D. S. (1994). Environmental Change in Grasslands - Assessment Using Models. *Climatic Change*, *28*(1-2), 111–141.

Parton, W. J., Ojima, D. S., & Schimel, D. S. (1994). ENVIRONMENTAL-CHANGE IN GRASSLANDS - ASSESSMENT USING MODELS. *Climatic Change*, *28*(1-2), 111–141.

Pascual, M., & Bouma, M. J. (2009). Do rising temperatures matter? *Ecology*, *90*, 906–931.

Pate, R. (2011). Assessment of algae biofuels resource demand and -up implications for the U.S. Department of energy of United States of America. *Algae Industry Magazine*, 29–31.

Peck, J. C. (2007). Groundwater Management in the High Plains Aquifer in the USA: Legal Problems and Innovations. In *The Agricultural Groundwater Revolution: Opportunities and Threats to Development* (pp. 296–319). CAB International.

Pendall, E., Osanai, Y., Williams, A., & Hovenden, M. (2010). Soil carbon storage under simulated climate change is mediated by plant functional type. *Global Change Biology*, *17*, 505–514.

Pender, J., Marré, A., & Reeder, R. (2012). *Rural Wealth Creation: Concepts, Strategies and Measures. ERR-131* (p. March).

Perkin, J. S., & Bonner, T. H. (2010). Long-term changes in flow regime and fish assemblage composition in the Guadalupe and San Marcos rivers of Texas. *River Research and Applications*, *27*(5), 566–579.

Perkin, J. S., & Gido, K. B. (2011). Stream Fragmentation Thresholds for a Reproductive Guild of Great Plains Fishes. *Fisheries*, *36*, 371–383.

Peterson, G. A., & Cole, C. V. (1995). Productivity of Great Plains Soils: Past, Present, and future. In S. R. Johnson & A. Bouzaher (Eds.), *Conservation of Great Plains Ecosystems* (pp. 325–342). Dordrecht, Netherlands: Kluwer Academic Publishers.

Peterson, G. A., & Westfall, D. G. (2004). Managing precipitation use in sustainable dryland agroecosystems. *Annals of Applied Biology*, *144*(2), 127–138.

Peterson, G. D., Cumming, G. S., & Carpenter, S. R. (2003). Scenario planning: a tool for conservation in an uncertain world. *Conservation Biology*, *17*(2), 358–366.

Pew Center on Global Climate Change. (2012). *Renewable & Alternative Energy Portfolio Standards*.

Phillips, W. a., Horn, G. W., & Cole, N. a. (2011). The Relevancy of Forage Quality to Beef Production. *Crop Science*, *51*(2), 410. doi:10.2135/cropsci2010.06.0382

Pidgeon, N., & Fischhoff, B. (2011). The role of social and decision sciences in communicating uncertain climate risks. *Nature Cliamte Change*, *1*, 35–41.

Pielke, S., & Roger, A. (2005). Minority Report, Comments Provided to the NRC Review Committee of the U.S. Climate Change Science Program's Synthesis and Assessment Product on Temperature Trends in the Lower Atmosphere. In *Atmospheric Science Bluebook No. 758* (p. 8). Fort Collins, CO: Colorado State University.

Pieper, R. D. (2005). Grasslands of central North America. In J. M. Suttie, S. G. Reynolds, & C. Batello (Eds.), *Grasslands of the World. Plant Production and Protection Series No 34* (p. 34). Rome, Italy: Food and Agriculture Organization, United Nations.

Pigg, J. (1987). Survey of fishes in the Oklahoma Panhandle and Harper County, northwestern Oklahoma. *Proceedings of the National Academy of Sciences*, *67*, 45–59.

Pigg, J. (1991). Decreasing distribution and current status of the Arkansas River Shiner, Notropis girardi, in the rivers of Oklahoma and Kansas. *Proceedings of the National Academy of Sciences, 71*, 5–15.

Platania, S. ., & Altenbach, C. S. (1998). Reproductive strategies and egg types of seven Rio Grande basin cyprinids. *Copeia, 1998*, 559–569.

Poff, N. L., Allan, J. D., Bain, M. B., Karr, J. R., Prestegaard, K. L., Richter, B. D., & Sparks, R. E. (1997). The natural flow regime: a paradigm for river conservation and restoration. *BioScience, 47*, 769–784.

Poff, N. L., & Zimmerman, J. K. H. (2010). Ecological responses to altered flow regimes: a literature review to inform the science and management of environmental flows. *Journal of Freshwater Biology, 55*, 194–205.

Polley, H. W. (1997). Implications of rising atmospheric carbon dioxide concentration for rangelands. *Journal of Range Management, 50*, 561–577.

Polley, H. W., Emmerich, W., Bradford, J. A., Sims, P. L., Johnson, D. A., Saliendra, N. Z., … Dugas, W. A. (2010). Precipitation regulates the response of net ecosystem CO2 exchange to environmental variation on United States rangelands. *Rangeland Ecology & Management, 63*, 176–186.

Polley, H. W., Jin, V. L., & Fay, P. A. (2011). CO2-caused change in plant species composition rivals the shift in vegetation between mid-grass and tallgrass prairies. *Global Change Biology, 18*(2), 700–710.

Poorter, H., & Navas, M. L. (2003). Plant growth and competition at elevated CO2: on winners, losers and functional groups. *New Phytologist2, 157*, 175–198.

Pope, J. (2007). *Conservation Takes Pressure Off Pumpkin CreekRep* (p. 3). Lincoln, Nebraska.

Potter, K. N., Jones, O. R., Torbert, H. A., & Unger, P. W. (1997). Crop rotation and tillage effects on organic carbon sequestration in the semiarid southern Great Plains. *Soil Science, 162*(2), 140–147.

Preston, B., Westaway, R., & Yuen, E. (2011). Climate adaptation planning in practice: an evaluation of adaptation plans from three developed nations. *Mitigation and Adaptation Strategies for Global Change, 16*(4), 407–438.

Propst, D. L., & Gido, K. B. (2004). Response of native and nonnative fishes to natural flow regime mimicry in the San Juan River. *Transactions of the American Fisheries Society, 133*, 922–931.

Pulwarty, R. (2003). Climate and water in the West: Science, information and decision-making. *Water Resources, 124*, 4–12.

Rabalais, N. N., Turner, R. E., & Wiseman, W. J. (2002). Gulf of Mexico Hypoxia, a.K.a. "the Dead Zone." *Annual Review of Ecology and Systematics, 33*(1), 235–263. doi:10.1146/annurev.ecolsys.33.010802.150513

Randolph, S. E. (2009). Perspectives on climate change impacts on infectious diseases. *Ecology, 90*, 927–931.

Ratajczak, Z., Nippert, J. B., Hartman, J. C., & Ocheltree, T. W. (2011). Positive feedbacks amplify rates of woody encroachment in mesic tallgrass prairie. *Ecosphere, 2*(11), 121.

Raun, L. (2010). *Trend Analysis of Ozone Concentrations in the City of Houston and Vicinity (2005-2009). April* (p. 16).

Reich, P. B., Tilman, D., Craine, J., Ellsworth, D., Tjoelker, M., & Mark, G. et al. (2001). Do species and functional groups differ in acquisition and use of C, N and water under varying atmospheric CO2 and N availability regimes? A field test with 16 grassland species. *New Phytologist, 150*, 435–448.

Renewable Fuels Association. (2011). Renewable Fuels Association. Statistics. Retrieved from www.ethanolrfa.org/pages/statistics

Renn, O., Klinke, A., & van Asselt, M. (2011). Coping with Complexity, Uncertainty and Ambiguity in Risk Governance: A Synthesis. *Ambio, 40*(2), 231–246.

Revenga, C., Campbell, I., Abell, R., De Villiers, P., & Bryer, M. (2005). Prospects for monitoring freshwater ecosystems towards the 2010 targets. *Philosophical Transactions of the Royal Society of London Series B, 360,* 397–413.

Reynolds, C., Crompton, L., & Mills, J. (2010). Livestock and climate change impacts in the developing world. *Outlook on Agriculture, 39*(4), 245–248.

Reynolds, R., Belnap, J., Reheis, M., Lamothe, P., & Luiszer, F. (2001). Aeolian dust in Colorado Plateau soils: nutrient inputs and recent change in source. *Proceedings of the National Academy of Sciences, 98,* 7123–7127.

Ribaudo, M. (2011). *Reducing agricultures nitrogen footprint: Are new policy approaches needed?. Amber Waves.*

Ribaudo, M., Delgado, J., Hansen, L., Livingston, M., Mosheim, R., & J., W. (2011). *Nitrogen In Agricultural Systems: Implications For Conservation Policy. ERR-127* (p. September).

Rice, C. W., Todd, T. C., Blair, J. M., Seastedt, T. R., Ramundo, R. A., & Wilson, G. W. T. (1998). Belowground Biology and Processes. In A. K. Knapp, J. M. Briggs, D. C. Hartnett, & S. L. Collins (Eds.), *Grassland Dynamics: Long-term ecological research in tallgrass prairie* (pp. 244–264). New York, NY: Oxford Press.

Richter, B. D., Baumgartner, J. V., Powell, J., & Braun, D. P. (1996). A method for assessing hydrologic alteration within ecosystems. *Conservation Biology, 10,* 1163–1174.

Roberts, B. (2011). *Potential for Photovoltaic Solar Installation in Non-Irrigated Corners of Center Pivot Irrigation Fields in the State of Colorado: Technical Report NREL/TP-6A20-51330* (p. 18). Golden, CO.

Robinson, L. W., & Berkes, F. (2011). Multi-level participation for building adaptive capacity: formal agency - community interactions in northern Kenya. *Global Environmental Change, 21,* 1185–1194.

Rogers, C. L. (2007). *The Sanitation Facilities Construction Program of the Indian Health Service Annual Report for 2007.*

Rogers, D. J., & Randolph, S. E. (2000). The global spread of malaria in a future, warmer world. *Science, 289,* 1763–1766.

Rosebud Sioux Tribe. (2012). Rosebud Sioux Reservation website.

Rosenzweig, C., Iglesias, A., Yang, X. B., Epstein, P. R., & Chivian, E. (2000). *Climate Change and U.S. Agriculture: The Impacts of Warming and Extreme Weather Events on Productivity, Plant Diseases, and Pests.*

Rosenzweig, C., Strzepek, K. M., Major, D. C., Iglesias, A., Yates, D. N., McCluskey, A., & Hillel, D. (2004). Water resources for agriculture in a changing climate: International case studies. *Global Environmental Change, 14,* 345–360.

Roth, D. (2010). Texas Hurricane History. Retrieved from http://www.hpc.ncep.noaa.gov/research/txhur.pdf

Roy, S., Chen, B. L., Girvetz, E., Maurer, E. P., Mills, W. B., & Grieb, T. M. (2010). *Evaluating sustainability of projected water demands under future climate change scenarios.* New York, NY.

Royal Academy of Engineering. (2011). *Infrastructure, Engineering, and Climate Change Adaptation – ensuring services in an uncertain future* (p. February). London, UK: Royal Academy of Engineering.

Sabo, J. L., Sinha, T., Bowling, L. C., Schoups, G. H. W., Wallender, W. W., Campana, M. E., ... Wohl, E. E. (2010). Reclaiming freshwater sustainability in the Cadillac Desert. *Proceedings of the National Academy of Sciences, 107*(21263-21270).

Sainju, U. M., Lenssen, A., Caesar-Thonthat, T., & Waddell, J. (2006). Carbon sequestration in dryland soils and plant residue as influenced by tillage and crop rotation. *Journal of Environmental Quality, 35*(4), 141–1347.

Sainju, U. M., Lenssen, A. W., Caesar-TonThat, T., Jabro, J. D., Lartey, R. T., Evans, R. G., & Allen, B. L. (2011). Dryland residue and soil organic matter as influenced by tillage, crop rotation, and cultural practice. *Plant and Soil, 338*(1), 27–41.

Sala, O. E., Parton, W. J., Joyce, L. A., & Lauenroth, W. K. (1988). Primary production of the central grassland region of the United States. *Ecology, 69*, 40–45.

Samson, F. B., Knopf, F. ., & Ostlie, W. R. (2004). Great Plains ecosystems: past, present, and future. *Wildlife Society Bulletin, 32*, 6–15.

Samson, F. B., & Knopf, F. L. (1994). Prairie conservation in North America. *BioScience, 44*, 418–421.

Samways, M. J. (2008). Dragonflies as focal organisms in contemporary conservation biology. In A. Córdoba-Aguilar (Ed.), *Dragonflies & Damselflies: Model Organisms for Ecological and Evolutionary Research* (pp. 97–108). New York, NY: Oxford University Press.

San Antonio Water System. (2009). Water Management Plan Update Fact Sheet. Retrieved January 03, 2012, from http://www.saws.org/our_water/WaterResources/2009wmp/docs/2009SAWSWaterManagementPlanFact.pdf

Savonis, M. J., Burkett, V. R., & Potter, J. R. (2008). *Impacts of climate change and variability on transportation systems and infrastructure: Gulf coast study, part i. Synthesis and assessment product 4.7.*

Scanlon, B. R., Reedy, R. C., & Gates, J. B. (2010). Effects of irrigated agroecosystems on (1) Quantity of soil water and groundwater in the southern High Plains, Texas. *Water Resources Research, 46*, W09537.

Scheffer, M., & Westley, F. R. (2007). The evolutionary basis of rigidity: Locks in cells, minds, and society. *Ecology and Society, 12*.

Schimel, D., Keller, M., & Duffy, P. (2009). *NEON Observatory Design Strategy.*

Schlosser, I. J. (1982). Fish community structure and function along two habitat gradients in a headwater stream. *Ecol. Monogr., 52*(4), 395–414.

Schmidt, K. A., & Ostfeld, R. S. (2001). Biodiversity and the dilution effect in disease ecology. *Ecological Society of America, 82*(3), 609–619.

Schmidt, T. L., & Leatherberry, E. C. (1995). Expansion of eastern redcedar in the Lower Midwest. *North J. Appl. For., 12*(4), 180–183.

Schneekloth, J., & Andales, A. (2009). *Seasonal Water Needs and Opportunities for Limited Irrigation for Colorado Crops. Fact sheet 4.718.* Fort Collins, CO. Retrieved from http://www.ext.colostate.edu/pubs/crops/04718.pdf

Schubert, S. D., Suarez, M. J., Pegion, P. J., Koster, R. D., & Bacmeister, J. T. (2004). On the cause of the 1930s Dust Bowl. *Science, 303*(5665), 1855–1859.

Schwartz, M. D., Ahas, R., & Aasa, A. (2006). Onset of spring starting earlier across the Northern Hemisphere. *Global Change Biology, 12*, 343–351.

Science Team about Energy and Prairie Pothole Environments. (2011). *Brine Contamination to Prairie Potholes from Energy Development in the Williston Basin.* Retrieved from http://steppe.cr.usgs.gov

Seastedt, T. R., Briggs, J. M., & Gibson, D. J. (1991). Controls of nitrogen limtation in tallgrass prairie. *Oecologia, 87*, 72–79.

Sellers, P. J. et al. (1996). Comparison of radiative and physiological effects of doubled atmospheric CO2 on climate. *Science, 271*(5254), 1402–1406.

Shapley, M., Ito, E., & Donovan, J. J. (2005). Authigenic calcium carbonate flux in groundwater-controlled lakes: implications for lacustrine paleoclimate records. *Geochim. Cosmochim. Acta, 69*, 2517–2533.

Sherry, R. A., Zhou, X., Gu, S., Arnone III, J. A., Schimel, D. S., Verburg, P. S., … Luo, Y. (2007). Divergence of reproductive phenology under climate warming. *Proceedings of the National Academy of Sciences, 104*, 198–202.

Simberloff, D. (2008). Invasion Biologists and the Biofuels Boom: Cassandras or Colleagues. *Weed Science, 56*, 867–872.

Sinervo, B., Méndez-de-la-Cruz, F., Miles, D. B., Heulin, B., Bastianns, E., & Villagrán-Santa Cruz, M. (2010). Erosion of lizard diversity by climate change and altered thermal niches. *Science, 328*, 894–899.

Skagen, S. K., Granfors, D. A., & Melcher, C. P. (2008). On determining the significance of ephemeral continental wetlands to North American migratory shorebirds. *Auk, 125*, 20–29.

Skagen, S. K., & Melcher, C. P. (2011). *Avian conservation in the Prairie Pothole Region, Northern Great Plains—Understanding the links between climate, ecosystem processes, wetland management, and bird communities: U.S. Geological Survey Fact Sheet 2011–3030* (p. 4).

Smit, B., Burton, I., Klein, R. J. T., & Wandel, J. (2000). An Anatomy of Adaptation to Climate Change and Variability. *Climatic Change, 45*, 223–251.

Smit, B., & Wandel, J. (2006). Adaptation, adaptive capacity and vulnerability. *Global Environmental Change, 16*(3), 282–292. doi:10.1016/j.gloenvcha.2006.03.008

Smith, L. M. (2003). *Playas of the Great Plains*. Austin, TX: University of Texas Press.

Smith, L. M., Haukos, D. A., McMurry, S. T., LaGrange, T., & Willis, D. (2011). Ecosystem services provided by playas in the High Plains: potential influences of USDA conservation programs. *Ecological Applications, 21*, S82–S92.

Smithers, J., & Smit, B. (1997). Human adaptation to climatic variability and change. *Global Environmental Change, 7*(2), 129–146.

Soussane, J. F., Tallec, T., & Blanford, V. (2010). Mitigating the greenhouse gas balance of ruminant production system through carbon sequestration in grasslands. *Animal, 4*, 334–350.

Spears, M., Harrison, A., Sankovich, V., Soddell, J., & Brekke, L. (2011). *Bureau of Reclamation Technical Service Center Water Resources Planning and Operations Support Group. Literature Synthesis on Climate Change Implications for Reclamation's Water Resources* (p. January).

Squires, J., & Korete, D. (2009). The Nuestra Casa Home Improvement Loan Program and the Old Dream of a Concrete Home. *Rural Voices, 14*(1), 9–10.

St. Pierre, N. R., Cobanov, B., & Schnitkey, G. (2003). Economic losses from heat stress by U.S. livestock industries. *J. Dairy Sci., 86*, E52–E77.

Stam, J. M., & Dixon, B. L. (2004). *Farmer Bankruptcies and Farm Exits in the United States, 1899–2002. Agriculture Bulletin No. 788*. Washington, DC, USA.

State of California Department of Water Resources. (2008). *Managing an Uncertain Future – Climate Change Adaptation Strategies for California's Water* (p. 30). October.

Statewide Water Quality Management Plan. (2011). *Arkansas River Basin Plan. Final Draft – June 1, 2011*. Retrieved from http://www.colorado.gov/cdphe/cse-google-static/?q=arkansas&cof=FORIDA10&ie=UTF-8&sa=Search

Staver, L. W., Staver, K. W., & Stevenson, J. C. (1996). Nutrient inputs to the Choptank River estuary: Implications for watershed management. *Estuaries, 19*(2B), 342–358.

Steyaert, P., & Jiggins, J. (2007). Governance of complex environmental situations through social learning: a synthesis of SLIM's lessons for research, policy and practice. *Environ. Sci. Policy, 10*(6), 575–586.

Stone, M. C., Hotchkiss, R. H., & Mearns, L. O. (2003). Water yield responses to high and low spatial resolution climate change scenarios in the Missouri River Basin. *Geophys. Res. Lett., 30*.

Strassberg, G., Scanlon, B. R., & Chambers, D. (2009). Evaluation of groundwater storage monitoring with the GRACE satellite: Case study of the High Plains aquifer, central United States. *Water Resources Research, 45*, W05410.

Struck, S. D., Jacobs, T., Moore, G., Pitt, R., Ports, M. A., O'Bannon, D., … Field, R. (2009). Kansas City – Balancing Green Infrastructure with Traditional Approaches for CSO Control. In *33rd International Association of Hydraulic Engineering and Research Congress: Water for a Sustainable Environment*.

Struthers, R., Hodge, F. S., Geishirt-Cantrell, B., & De Cora, L. (2003). Participant Experiences of Talking Circles on Type 2 Diabetes in Two Northern Plains American Indian Tribes. *Qualitative Health Research, 13*(8), 1094–115.

Stuart, A. J. (2008). Mammalian extinctions in the late Pleistocene of Northern Eurasia and North America. *Biological Reviews, 66*, 453–562.

Swaddle, J. P., & Calos, S. E. (2008). Increased avian diversity is associated with lower incidence of human West Nile infection: observation of the dilution effect. *PloS One, 3*, e2488.

Swanson, F. J., Kratz, T. K., Craine, N., & Woodmanser, R. G. (1988). Landform effects on ecosystem patterns and process. *BioScience, 3B*, 92–96.

Sylvester, K. M., & Cunfer, G. P. (2009). An Unremembered Diversity: Mixed Husbandry and the American Grasslands. *Agricultural History, 83*(3), 352–383.

Tanaka, J. A., Brunson, M., & Torell, L. A. (2011). A social and economic assessment of rangeland conservation practices. In D. D. Briske (Ed.), *Conservation Benefits of Rangeland Practices: Assessment, Recommendations and Knowledge Gaps* (pp. 371–422). Washington, DC, USA: USDA Natural Resources Conservation Service.

Tangen, B. A., Butler, M. G., & Ell, M. J. (2003). Weak correspondence between macroinvertebrate assemblages and land use in prairie pothole region wetlands, USA. *Wetlands, 23*(1), 104–115.

Taylor, C. M. (2010). Covariation among plains stream fish assemblages, flow regimes, and patterns of water use. In K. B. Gido & D. A. Jackson (Eds.), *Community Ecology of Stream Fishes: Concepts, Approaches and Techniques* (pp. 447–459). Bethesda, M.D.

Taylor, C. M., Millican, D. S., Roberts, M. E., & Slack, W. T. (2008). Long-term changes to fish assemblages and the flow regime in a southwestern U.S. river system after extensive aquatic ecosystem fragmentation. *Ecography, 31*, 787–797.

Taylor, L. H., Latham, S. M., & Woolhouse, M. E. J. (2001). Risk factors for human disease emergence. *Philos Trans R Soc Lond B Biol Sci, 356*, 983–989.

Texas A&M University. (2012). Agrilife. Retrieved January 03, 2012, from http://agrilife.org

Texas Water Development Board. (2011). *History of Water Reuse in Texas*.

The Heinz Center. (2008a). *Strategies for Managing the Effects of Climate Change on Wildlife and Ecosystems* (p. 43). Washington, DC.

The Heinz Center. (2008b). *The state of the nation's ecosystems 2008: measuring the land, waters, and living resources of the United States*. Washington, DC, USA: Island Press.

The Heinz Center. (2009). *Measuring the Results of Wildlife Conservation Activities* (p. 122). Washington, DC, USA.

Thomas, D. S. G., & Twyman, C. (2005). Equity and justice in climate change adaptation amongst natural-resource-dependent societies. *Glob. Environ. Change-Human Policy Dimens., 15*(2), 115–124.

Thompson, J. (2010). *Workshop reports: Border Crossing: Preparing for & Adapting to Climate Change Effects in Northern Colorado* (pp. November, 2010).

Thompson, M., Ellis, R., & Wildavsky, A. (1990). *Cultural Theory*. Boulder, CO: Westview Press.

Thomson, A. M., Brown, R. A., Rosenberg, N. J., Srinivasan, R., & Izaurralde, C. (2005). Climate change impacts for the conterminous USA: an integrated assessment. *Climatic Change, 69*, 67–88.

Tidwell, V., Cha-tien Sun, A., & Malczynski, L. (2011). *Biofuel Impacts on Water. Sandia Report. SAND2011-0168.*

Tieszen, L. L., Reed, B. C., Bliss, N. B., Wylie, B. K., & DeJong, D. D. (1997). NDVI, C3 and C4 production, and distributions in Great Plains grassland land cover classes. *Ecological Applications, 7*(1), 59–78.

Torell, L. A., Murugan, S., & Ramirez, O. A. (2010). Economics of flexible versus conservative stocking strategies to manage climate variability risk. *Rangeland Ecology & Management, 63*, 415–425.

Trinity River Authority of Texas. (2010). *Trinity River Basin Master Plan*. Retrieved from http://www.trinityra.org/downloads/Master Plan Justified REV.pdf

Tronstad, R., & Feuz, D. (2002). Impacts of the 2002 Drought on Western Ranches and Public Land Policies. *Western Economics Forum, 1*(2), 19–23.

Tsai, J.-S., Venne, L. S., McMurry, S. T., & Smith, L. M. (2007). Influences of land use and wetland haracteristics on water loss rates and hydroperiods of playas in the Southern High Plains, USA. *Wetlands, 27*(3), 683–692.

Tsai, J.-S., Venne, L. S., McMurry, S. T., & Smith, L. M. (2010). Vegetation and land use impact on water loss rate in playas of the Southern High Plains, USA. *Wetlands, 30*, 1107–1116.

Tschakert, P., Coomes, O. T., & Potvin, C. (2007). Indigenous livelihoods, slash-and-burn agriculture, and carbon stocks in Eastern Panama. *Ecol. Econ., 60*, 807–820.

Turka, R. J., & Gray, R. E. (2005). Impacts of coal mining. *Reviews in Engineering Geology, 16*, 79–86.

Turner, B. L., Matson, P. a, McCarthy, J. J., Corell, R. W., Christensen, L., Eckley, N., … Tyler, N. (2003). Illustrating the coupled human-environment system for vulnerability analysis: three case studies. *Proceedings of the National Academy of Sciences of the United States of America, 100*(14), 8080–5. doi:10.1073/pnas.1231334100

Tzoulas, K., Korpela, K., Venn, S., Yli-Pelkonen, V., Kazmierczak, A., Niemela, J., & James, P. (2007). Promoting Ecosystem and Human Health in Urban Areas Using Green Infrastructure: A Literature Review. *Landscape and Urban Planning, 81*, 167–178.

U.S. Army Corps of Engineers. (1998). *Missouri River Main Stem Reservoirs System Description and Operation. Reservoir Control Center, Northwestern Division – Missouri River Basin. 1998* (p. Fall 1998).

U.S. Army Corps of Engineers. (2002). *Mississippi River and tributaries, Atchafalaya Basin, Louisiana, Lower Atchafalaya Basin reevaluation study*. Vicksburg, MS.

U.S. Army Corps of Engineers. (2006). *Missouri River Mainstem Reservoir System Master Water Control Manual Missouri River Basin. Reservoir Control Center, Northwestern Division – Missouri*

River Basin. Omaha, Nebraska. Retrieved from http://www.nwd-mr.usace.army.mil/rcc/reports/mmanual/MasterManual.pdf

U.S. Army Corps of Engineers. (2009). *Western States Watershed Study: Report to the Western States Water CouncilRep* (p. 42).

U.S. Army Corps of Engineers. (2010). *Storage Reallocation Report Lake Texoma Oklahoma and Texas. Tulsa District* (p. March 2010). Retrieved from http://www.swt.usace.army.mil/library/Lake Texoma Reallocation Study/MARCH_2010_FINAL_With_Addendum/01 Main and App A thru G Texoma MARCH 2010 FINAL and Addendum.pdf

U.S. Army Corps of Engineers. (2012). *What is the McClellan-Kerr Arkansas River Navigation System (MKARNS)*. Retrieved from http://www.swl.usace.army.mil/navigation/mckarns.html

U.S. Bureau of Reclamation. (1996). *Wind River Indian Reservation – Municipal, Rural, and Industrial Water Supply Needs Assessment – Draft Report* (p. 130).

U.S. Bureau of Reclamation. (2008a). *Southern Delivery System Final Environmental Impact Statement* (p. Dece,ber 2008).

U.S. Bureau of Reclamation. (2008b). *Yuma Desalting Plant, Demonstration Run Report*.

U.S. Bureau of Reclamation. (2010). *Annual Operating Plan: Fryingpan-Arkansas Project. Water Year 2010 Operations*. Retrieved from http://www.usbr.gov/gp/aop/fa/fa10.pdf

U.S. Bureau of Reclamation. (2011). *SECURE Water Act Section 9503(c) – Reclamation Climate Change and Water, Report to Congress. 226*. Retrieved from http://www.usbr.gov/climate/SECURE/docs/SECUREWaterReport.pdf

U.S. Bureau of Reclamation. (2012a). *Colorado-Big Thompson Project*. Retrieved from http://www.usbr.gov/projects/Project.jsp?proj_Name=Colorado-Big+Thompson+Project

U.S. Bureau of Reclamation. (2012b). *Projects and Facilities Database Reclamation's Portal for Information on Dams, Powerplants and Projects*. Retrieved from http://www.usbr.gov/projects/

U.S. Census Bureau. (2000). 2000 Census Data.

U.S. Census Bureau. (2010a). 2010 Census Data: Population size. U.S. Bureau of the Census. Retrieved from http://2010.census.gov/2010census/data/

U.S. Census Bureau. (2010b). *American Community Survey. 2005-2009*. Retrieved from http://www.census.gov/acs/www/

U.S. Census Bureau. (2011). *Population Distribution and Change: 2000 to 2010. 2010 Census Briefs* (p. March 2011).

U.S. Climate Change Science Program. (2008a). *Abrupt Climate Change. A report by the U.S. Climate Change Science Program and the Subcommittee on Global Change Research*. (P. U. Clark, A. J. (coordinating lead authors) Weaver, E. Brook, E. R. Cook, T. L. Delworth, & K. (chapter lead authors) Steffen, Eds.) (p. 244). Reston, VA.

U.S. Climate Change Science Program. (2008b). *The effects of climate change on agriculture, land resources, water resources, and biodiversity in the United States. A Report by the U.S. Climate Change Science Program and the Subcommittee on Global Change Research*. (P. Backlund, A. Janetos, D. Schimel, J. Hatfield, K. Boote, P. Fay, … R. Shaw, Eds.) (p. 362). Washington, DC, USA.

U.S. Climate Change Science Program. (2009). *Thresholds of Climate Change in Ecosystems. A Report by the U.S. Climate Change Science Program and the Subcommittee on Global Change Research*. (D. B. Fagre, C. W. Charles, C. D. Allen, C. Birkeland, F. S. Chapin, P. M. Groffman, … G. Sugihara, Eds.). Reston, VA.

U.S. Commission on Civil Rights. (2003). *Federal Funding and Unmet Needs In Indian Country*.

U.S. Department of Energy. (2006). *Energy Demands on Water Resources: Report to Congress on the Interdependency of Energy and Water* (p. 80). Retrieved from http://www.sandia.gov/energy-water/docs/121-RptToCongress-EWwEIAcomments-FINAL.pdf

U.S. Department of Energy. (2009). 2009 Energy Consumption Per Person. Retrieved January 03, 2012, from http://energy.gov/maps/2009-energy-consumption-person

U.S. Department of Housing and Urban Development. (1996). *Assessment of American Indian Housing Needs.*

U.S. Environmental Protection Agency. (2008a). Ensuring a Sustainable Future: An Energy Management Guidebook for Wastewater and Water Utilities. Retrieved from http://www.epa.gov/owm/waterinfrastructure/pdfs/guidebook_si_energymanagement.pdf

U.S. Environmental Protection Agency. (2008b). *Inventory of U.S. Greenhouse gas emissions and sinks: 1990-2006. EPA 430-R-08-005. April 15, 2008*. Washington, DC, USA. Retrieved from http://www.epa.gov/climatechange/emissions/downloads/08_CR.pdf

U.S. Environmental Protection Agency. (2009). *Office of Water. Drinking Water Infrastructure Needs Survey and Assessment – Fourth Report to Congress. February 2009* (p. 64).

U.S. Environmental Protection Agency. (2010a). *Green Infrastructure in Arid and Semi-Arid Climates* (p. May 2010).

U.S. Environmental Protection Agency. (2010b). *Society and Ecosystems in Climate Change Indicators in the United States. EPA 430-R-10-007.*

U.S. Environmental Protection Agency. (2011). *Drinking Water Infrastructure Needs Survey and Assessment. EPA Form 6100-01. OMB No.: 2040-0274* (p. March 2011).

U.S. Environmental Protection Agency. (2012a). Cutting Energy Usage & Costs. *Water: Sustainable Infrastructure.* Retrieved from http://water.epa.gov/infrastructure/sustain/cut_energy.cfm

U.S. Environmental Protection Agency. (2012b). *Particulate Matter Basic Information.* Retrieved from http://www.epa.gov/pm/basic.html

U.S. Environmental Protection Agency. (2012c). State Funding Sources and Renewable Portfolio Standards. Retrieved from http://www.epa.gov/lmop/publications-tools/funding-guide/state-resources/index.html

U.S. Environmental Protection Agency. (2012d). Sustainable Infrastructure for Water Programs. *EPA Region 7.* Retrieved from http://www.epa.gov/region7/water/si.htm

U.S. Environmental Protection Agency. (2012e). Water & Energy Efficiency. *Water: Sustainable Infrastructure.* Retrieved from http://water.epa.gov/infrastructure/sustain/waterefficiency.cfm

U.S. Fish and Wildlife Service. (2005). *The Endangered Species Program.* Retrieved from www.USFWS.gov/endangered/

U.S. Geological Survey. (1990). *Water Fact Sheet. Largest Rivers in the United States. Open-File Report 87-242.* Retrieved from http://pubs.usgs.gov/of/1987/ofr87-242/pdf/ofr87242.pdf

U.S. Geological Survey. (2003). *Coal-Bed Gas Resources of the Rocky Mountain Region. National Assessment of Oil and Gas Fact Sheet.*

U.S. Geological Survey. (2005). *Estimated Use of Water in the United States, County-Level Data for 2005.* Retrieved from http://water.usgs.gov/watuse/data/2005/

U.S. Geological Survey. (2009). *Estimated Use of Water in the United States County-Level Data for 2005.* Retrieved from http://water.usgs.gov/wsi/

Universitiy of Nebraska. (2011). *Nebraska Community Forestry Conference and Tree City USA Recognition Day report.*

USDA. (2011). *Agriculture and Forestry Greenhouse Gas Inventory: 1990-2008. Technical Bulletin 1930.* Retrieved from www.usda.gov/oce/climate_change/AFGGInventory1990_2008.htm

USDA. (2012). USDA Census of Agriculture. Historical Census Publications. *U.S. Department of Agriculture*. Retrieved January 01, 2012, from http://www.agcensus.usda.gov/Publications/Historical_Publications/index.php

USDA Economic Research Service. (2012a). *Amber Waves Data Feature: Mapping Food Deserts in the U.S.* Retrieved from http://www.ers.usda.gov/AmberWaves/December11/DataFeature/index.htm

USDA Economic Research Service. (2012b). Atlas of Rural and Small-town America. Retrieved from http://www.ers.usda.gov/data/ruralatlas/index.htm

USDA Economic Research Service. (2012c). *Data Sets: Rural definitions*. Retrieved from http://www.ers.usda.gov/Data/RuralDefinitions/

USDA Economic Research Service. (2012d). State Fact Sheets, Data Sets. Retrieved from http://www.ers.usda.gov/StateFacts/

USDA National Agricultural Statistics Service. (2007). 2007 Census of Agriculture. United States Summary and State Data. Vol. 1, Geographic area series, part 51, AC-07-A-51. Retrieved from www.222.agcensus.usda.gov

USDA National Agricultural Statistics Service. (2009). Agricultural Statistics. NASS Home Page. Retrieved January 03, 2012, from http://www.nass.usda.gov/

USDA National Agricultural Statistics Service. (2010). *Farm and Ranch Irrigation Survey 2008. Volume 3- Part 1, AC-07-SS-1. Updated 2010*. Retrieved from http://www.agcensus.usda.gov/Publications/2007/Online_Highlights/Farm_and_Ranch_Irrigation_Survey/index.asp

USDA Natural Resources Conservation Service. (1996). *State of the Land for the Northern Plains Region*. (p. 64). Lincoln, Nebraska.

Van de Meene, S. J., Brown, R. R., & Farrelly, M. A. (2011). Towards Understanding Governance for Sustainable Urban Water Management. *Global Environmental Change: Human and Policy Dimensions*, *21*(3), 1117–1127.

Van Dijk, J., Sargison, N. D., Kenyon, F., & Skuce, P. J. (2010). Climate change and infectious disease: helminthological challenges to farmed ruminants in temperate regions. *Animal*, *4*(3), 377–392.

Vasconcelos, J. T., & Galyean, M. L. (2007). Nutritional recommendations of feedlot consulting nutritionists: the 2007 Texas Tech University survey. *Journal of Animal Science*, *85*(10), 2772–81. doi:10.2527/jas.2007-0261

Vaswani, R., & Gasteyer, S. (2004). *Analyzing the Availability of Water and Sanitation Services in the United States*. Retrieved from http://win-water.org/reports/RCAP_full_final.pdf

Ver Ploeg, M., Breneman, V., Farrigan, T., Hamrick, K., Hopkins, D., Kaufman, P., … Tuckermanty, E. (2009). *Access to Affordable and Nutritious Food—Measuring and Understanding Food Deserts and Their Consequences: Report to Congress. Admin Pub No. (AP-036)* (p. 160).

Verweij, M., Douglas, M., Ellis, R., Engel, C., Hendriks, F., Lohmann, S., … Thompson, M. (2006). Clumsy solutions for a complex world: The case of climate change. *Public Adm*, *84*(4), 817–843.

Vining, K. C. (2002). *Simulation of Streamflow and Wetland Storage, Starkweather Coulee Subbasin, North Dakota, Water Years 1981–98. Water-Resources Investigations Report 02-4113* (p. 24). Bismark, ND.

Vitousek, P. M., Aber, J. D., Howarth, R. W., Likens, G. E., Matson, P., Schindler, D. W., … Tilman, G. D. (1997). Human alteration of the global nitrogen cycle: Sources and consequences. *Ecological Applications*, *7*, 737–750.

Vogel, C., & O'Brien, K. (2006). Who can eat information? Examining the effectiveness of season climate forecasts and regional climate-risk management strategies. *Climate Res., 33,* 111–112.

Walker, B., & Meyers, J. A. (2004). Thresholds in ecological and social-ecological systems: a developing data base. *Ecology and Society, 9*(2), 3.

Wan, S., Hui, D., Wallace, L., & Lou, Y. (2005). Direct and indirect effects of experimental warming on ecosystem carbon processes in a tallgrass prairie. *Global Biogeochemical Cycles, 19.*

Wand, S. J. E., Midgley, G. F., Jones, M. H., & Curtis, P. S. (1999). Responses of wild C4 and C3 grass (Poaceae) species to elevated atmospheric CO2 concentration: a test of current theories and perceptions. *Global Change Biology, 5,* 723–741.

Wedin, D. A., & Tilman, D. (1993). Competition among grasses along a nitrogen gradient - Initial conditions and mechanisms of competition. *Ecol. Monogr., 63,* 199–229.

Weiss, A., Hays, C. J., & Won, J. (2003). Assessing winter wheat responses to climate change scenarios: A simulation study in the US Great Plains. *Climatic Change, 58,* 119–147.

Weltzin, J. F., Loik, M. E., Schwinning, S., Williams, D. G., Fay, P. A., Haddad, B. M., … Zak, J. C. (2003). Assessing the response of terrestrial ecosystems to potential changes in precipitation. *BioScience, 53*(10), 941–952.

Western Area Power Administration. (2012). Western Area Power Administration (WAPA). *About the Upper Great Plains Regional Office.* Retrieved from http://www.wapa.gov/ugp/aboutus/default.htm

Western Governors' Association. (2006). *Water Needs and Strategies for a Sustainable Future* (p. 26).

Western Governors' Association. (2008). *Water Needs and Strategies for a Sustainable Future: Next StepsRep.* Denver, CO.

Western Governors' Association. (2010). *Climate Adaptation Priorities for the Western States: Scoping ReportRep.* (p. 20). Denver, CO.

Western States Water Council. (2011). *Western Water Resources Infrastructure Strategies: Identifying, Prioritizing and Financing Needs* (p. June).

Westfall, D. G., Peterson, G. A., & Hansen, N. C. (2010). Conserving and optimizing limited water for crop production. *Journal of Crop Improvement, 24*(1), 70–84.

Whitford, W. G., & Steinberger, Y. (2011). Effects of simulated storm sizes and nitrogen on three Chihuahuan Desert perennial herbs and a grass. *Journal of Arid Environments, 75,* 861–864.

Wilby, R. L., Troni, J., Biot, Y., Tedd, L., Hewitson, B. C., Smith, D. M., & Sutton, R. T. (2009). A review of climate risk information for adaptation and development planning. *International Journal of Climatology, 29*(9), 1193–1215.

Wilby, R. L., & Vaughan, K. (2011). Hallmarks of organisations that are adapting to climate change. *Water Environ. J., 25*(2), 271–281.

Wilcox, C. (2001). Habitat size and isolation affect colonization of seasonal wetlands by predatory aquatic insects. *Israeli Journal of Zoology, 47,* 459–475.

Wilde, G. R., & Durham, B. W. (2008). A life history model for peppered chub, a broadcast spawning cyprinid. *Transactions of the American Fisheries Society, 137,* 1657–1666.

Wilhelmi, O. V., & Wilhite, D. A. (2002). Assessing vulnerability to agricultural drought: A Nebraska case study. *Natural Hazards, 25,* 37–58.

Wilhite, D. A., & Pulwarty, R. S. (2005). Drought and water crises: lessons learned and the road ahead. In D. A. Wilhite (Ed.), *Drought and water crises: science, technology, and management issues* (pp. 389–398). Boca Raton, Florida: CRC Press.

Wilhite, D. A., Svoboda, M. D., & Hayes, M. J. (2007). Understanding the complex impacts of drought: A key to enhancing drought mitigation and preparedness. *Water Resources Management, 21*, 763–774.

Williams, D. D. (1997). Temporary ponds and their invertebrate communities. *Aquatic Conservation: Marine and Freshwater Ecosystems, 7*, 105–117.

Williams, D. D. (2006). *The Biology of Temporary Waters*. New York, NY: Oxford University Press.

Williams, M. A. (2007). Response of microbial communities to water stress in irrigated and drought-prone tallgrass prairie soils. *Soil Biology and Biochemistry, 39*, 2750–2757.

Willows, R. I., & Connell, R. K. (2003). *Climate adaptation: Risk, uncertainty and decision-making. UKCIP Technical Report*. Oxford, UK.

Wissel, B., Cooper, R. N., Leavitt, P. R., & Pham, S. V. (2011). Hierarchical regulation of pelagic invertebrates in lakes of the northern Great Plains: a novel model for interdecadal effects of future climate change on lakes. *Global Change Biology, 17*, 172–185.

Wolfe, N. D., Dunavan, C. P., & Diamond, J. (2007). Origins of major human infectious diseases. *Nature, 447*, 279–283.

Wolkovich, E. M., & Cleland, E. . (2011). The phenology of plant invasions: a community ecology perspective. *Front Ecol Environ, 9*(5), 287–294.

Wu, Y. P., Liu, S. G., & Abdul-Aziz, O. I. (2012). Hydrological effects of the increased CO(2) and climate change in the Upper Mississippi River Basin using a modified SWAT. *Climatic Change, 110*(3-4), 977–1003.

Xia, J., Niu, S., & Wan, S. (2009). Response of ecosystem carbon exchange to warming and nitrogen addition during two hydrologically contrasting growing seasons in a temperate steppe. *Global Change Biology, 15*, 1544–1556.

Zack, S., Ellison, K., Cross, M., & Rowland, E. (2010). *Climate Change Planning for the Great Plains: Wildlife Vulnerablity Assessment and Recommendations for Land and Grazing Management*. Bozeman, MT.

Zimmer, K. D., Hanson, M. A., & Butler, M. G. (2000). Factors influencing invertebrate communities in prairie wetlands: a multivariate approach. *Canadian Journal of Fisheries and Aquatic Science, 57*, 76–85.

Ziska, L. H. (2003). Evaluation of the growth response of six invasive species to past, present and future carbon dioxide concentrations. *Journal of Experimental Botony, 54*, 395–404.

Ziska, L. H., & George, K. (2004). Rising carbon dioxide and invasive, noxious plants: potential threats and consequences. *Water Resources Review, 16*, 427–46.

Ziska, L. H., & Runion, G. B. (2006). Fugure weed, pest and disease problems for plants. In P. Newton, A. Carran, G. Edwards, & P. Niklaus (Eds.), *Agroecosystems in a Changing Climate2*. New York, NY: CRC Press.

Ziska, L. H., Teasdale, J. R., & Bunce, J. A. (1999). Future atmospheric carbon dioxide concentrations may increase tolerance to glyphosate [N-(phosphonmethyl) glycine] in weedy species. *Weed Science1, 47*, 608–615.

Ziska, L., Knowlton, K., Rogers, C., Dalan, D., Tierney, N., Elder, M. A., … Frenz, D. (2011). Recent warming by latitude associated with increased length of ragweed pollen season in central North America. *Proceedings of the National Academy of Sciences, 108*, 4248–4251.

Editor Biographies

Dr. Dennis Ojima
Dr. Dennis Ojima is a Professor, Department of Ecosystem Science and Sustainability; Senior Research Scientist, Natural Resource Ecology Laboratory at Colorado State University; and University Director of the North Central Climate Science Center at Colorado State University for the Department of Interior. His research areas include climate and land use changes on ecosystems around the world, carbon accounting, and adaptation and mitigation strategies to climate change. He is Aldo Leopold Leadership Fellow, is serving on the National Research Council Board on Environmental Change and Society, and was Resident Senior Scholar at the H. John Heinz III Center for Science, Economics, and the Environment in Washington, DC. He has been recognized for his international contributions in the Millennium Ecosystem Assessment receiving which received the 2005 Zayed International Prize for the Environment and the International Panel on Climate Change (IPCC) 2007 Nobel Peace Prize. In 2013 was honored as a Champion of the Environment by the Mongolian Minister of the Environment and Green Development. Professor Ojima received his BA and Master's Degrees in Botany from Pomona College (1975) and the University of Florida (1978), and his PhD from the Rangeland Ecosystem Science Department at Colorado State University in 1987.

Dr. Jean Steiner
Dr. Steiner obtained a B.A. degree in geology from Cornell College, Mt. Vernon, Iowa; and M.S. and Ph.D. degrees in agronomy with a focus on agroclimatology from Kansas State University. She has worked for ARS since 1983, and is Director of the Grazinglands Research Laboratory in El Reno, OK, where she leads the Southern Plains Long Term Agroecosystem Research site. The laboratory also hosts the USDA Southern Plains Regional Climate Hub. Her personal research has spanned irrigated and rainfed cropping, forage-grazing systems, crop residue management, evapotranspiration, watershed management, and integrated systems research, addressing fundamental aspects of agroecological systems to quantify constraints to productivity and management impacts on environmental outcomes. Dr. Steiner is the 2014 President-Elect of the American Society of Agronomy and is Fellow of the American Society of Agronomy, the Soil and Water Conservation Society, the Soil Science Society of America, and the American Association for the Advancement of Sciences.

Dr. Shannon McNeeley
Dr. Shannon McNeeley received her doctoral degree in Environmental Change and Sustainability Science (ecological anthropology, ecology, climatology) from the University of Alaska Fairbanks (UAF) in the interdisciplinary Resilience and Adaptation Program as an NSF IGERT Fellow then as an NSF Graduate Research Fellow. Her doctoral research focused on climate variability and change impacts, vulnerabilities, and adaptive capacity of indigenous people (Athabascan Indians) in the remote, rural Interior region of Alaska. This was in close collaboration with tribes, state, and federal agency partners.

She first began working for the National Center for Atmospheric Research (NCAR) in 2000 as an associate scientist before starting her doctoral degree in the fall of 2004. Her work is interdisciplinary and cross-cultural incorporating the social and natural sciences in order to understand human-environment relationships and how people are impacted by and respond to environmental change. She has been involved in climate change education and research for over 16 years. As a postdoctoral fellow at NCAR, her research focused on water scarcity and sustainability in the context of climate variability and change and the Yampa/White Basins region of northwest Colorado. Then as a research fellow at the School of Natural Resources and Environment the University of Michigan, Dr. McNeeley co-wrote the Adaptation chapter of the upcoming U.S. National Climate Assessment and led research on climate adaptation actions implemented across the globe through the Global Environmental Facility financing mechanisms for developing and Least Developed Countries. She is currently a postdoctoral fellow at the DOI-sponsored North Central Climate Science Center at Colorado State University. In addition to continuing research on vulnerability and adaptation in water resource management, this will also entail working to build the capacity of the NC CSC to conduct and support regional assessment on climate change adaptive capacity and decision making.

Dr. Karen Cozzetto

Dr. Karen Cozzetto is a research hydrologist with the Western Water Assessment and the acting managing director of the Center for Water, Earth Science, and Technology at the Institute of Arctic and Alpine Research at the University of Colorado, Boulder. Her research interests include: surface water-groundwater interactions, climate influences on interannual streamflow variation, stream restoration, and working with resource managers on climate change adaptation planning. She was the lead investigator of a Native Communities and Climate Change Preparedness project and a contributing author to the Indigenous Peoples, Land, and Resources chapter of the third U.S. National Climate Assessment. Additionally, Dr. Cozzetto studies glacial meltwater streams in the McMurdo Dry Valleys of Antarctica as part of a Long-Term Ecological Research site. Dr. Cozzetto received her Ph.D. in Water Resources Engineering from the University of Colorado, Boulder in 2009.

Amber Childress

Amber Childress is a PhD student in the Graduate Degree Program in Ecology at Colorado State University. As a NSF IGERT Fellow in the IWATER (Integrated Water, Atmosphere, Ecosystems Education and Research) Program, she works in the Natural Resource Ecology Lab on climate change impacts and adaptation. Her research evaluates the capacity of water providers in the South Platte River Basin to adapt to future water stress, induced by climate, population, and land-use changes. Previously, she worked at the Heinz Center for Science, Economics and the Environment as part of the Terrestrial Carbon Group Secretariat on financial and policy mechanisms to achieve climate change mitigation from terrestrial ecosystems. Past experience also includes work in international project finance, focused in Central and South America and policy work in the office of the Speaker of the U.S. House of Representatives and within the UNFCCC. She holds a bachelor's degree in International Studies and Economics from Austin College and a master's degree from Colorado State University in Ecology.

Island Press | Board of Directors